Constellation Handbook

Constellation Handbook

Antonín Rükl

Sterling Publishing Co., Inc
New York

Library of Congress Cataloging-in-Publication Data Available

10 9 8 7 6 5 4 3 2 1

Published in 2003 by Sterling Publishing Co., Inc.
387 Park Avenue South, New York, NY 10016

Text, illustrations and graphic design by Antonín Rükl
Translated by Slavoš Kadečka

Originally published by Sterling Publishing Co., Inc.
under the title *Constellation Guidebook*

Distributed in Canada by Sterling Publishing
c/o Canadian Manda Group, One Atlantic Avenue, Suite 105
Toronto, Ontario, Canada M6K 3E7

Printed in the Czech Republic

Sterling ISBN 1-4027-0867-X
3/15/39/76-04

CONTENTS

Introduction 7

Where, When, and How to Find Constellations 9

The Zodiac and the Planets 16

A Celestial Guide 19

Deep Space Objects 20

Celestial Maps 32

Constellation Maps 46

Index of Constellations 216

Glossary of Terms 218

General Index 222

Advice, Hints and Instructions for Beginning Amateur Astronomers

Angular Measurements 50

Visible and Invisible Radiation 64

Observing Through the Atmosphere 74

Atmospheric Refraction 76

How Many Stars Can We See in the Sky? 88

Star Colors 90

Novae and Supernovae 92

Visitors to Constellations 96

Three-Dimensional View of the Universe 112

Observing with Binoculars 120

Messier's Objects 132

Why a Telescope? 148

Starry Sky in the Computer 170

Photographing the Starry Sky 174

Observation Records 178

Amateur Astronomy 200

Note:
The stellar distances as given in this book are based on the HIPPARCOS Catalogue (European Space Agency, 1997).

1. The constellation of Taurus (the Bull) in Hevelius' atlas of 1690.

INTRODUCTION

This is a guidebook to an unusual gallery. The gallery is open every day and there is no entrance fee. It belongs to all of us: it is our universe. And the rotating Earth offers the perceptive observer an ever changing panorama. The most impressive view of the universe is at night. The sky strewn with stars is like an amply illustrated book that we are learning to read. Our stellar ABC begins with the constellations.

The oldest constellations found on contemporary maps were created by the ancient Sumerians, Babylonians, and other inhabitants of the region between the Euphrates and the Tigris – Mesopotamia. The ancient Egyptians, and later the Greeks, added many other constellations.

Aratus of Soli, in his poem *Phaenomena,* presented a poetic picture of the constellations of Greek mythology in the 3rd century B.C. Its content formed the basis of the work of Ptolemy, the Greek astronomer, who compiled in about A.D. 150 a list of 48 constellations to which other astronomers have continued to add.

Since ancient times the constellations have been described and pictured as complex figures of animals, mythological beings, heroes, and symbols (Fig.1). The figures were not used merely for artistically rendering maps; they served as a necessary aid to identifying specific regions of the sky.

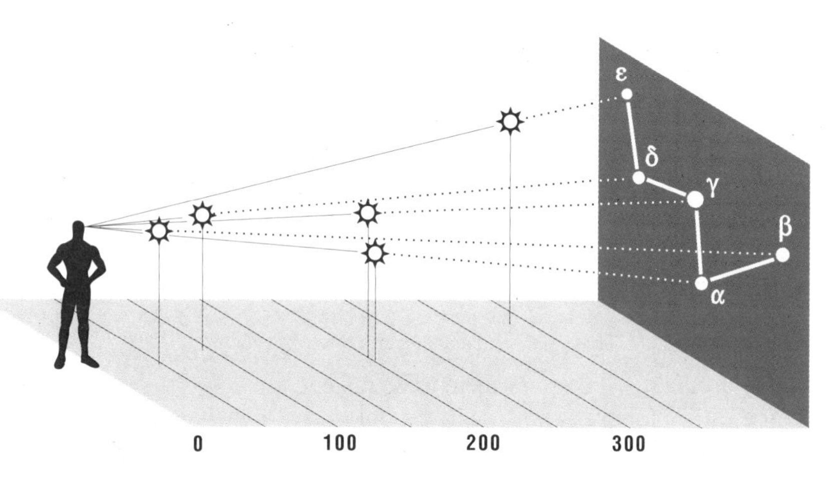

2. The constellations are formed by random groups of stars in the sky at different distances from Earth.

Toward the end of the 16th century, two Dutch navigators, Pieter Dirckszoon Keyser and Frederick de Houtman, proposed 12 new constellations in the southern sky that were adopted by Johann Bayer (1572–1625) and included in his classic atlas *Uranometria* in 1603. Additional constellations were described by Johannes Hevelius (1611–1687), an astronomer from Danzig, and by the French astronomer Nicolas-Louis de Lacaille (1713–1762), who introduced 14 more constellations named mostly after scientific instruments and art aids. As late as the 19th century cartographers continued to add on to or change the stellar gallery.

In 1930, the International Astronomical Union (IAU) divided the sky into 88 officially recognized constellations. Since that date, the constellations have been established as fixed components of the sky, bordered by accurately defined celestial parallels and meridians. The Latin names of the constellations and their abbreviations given in this book are internationally valid.

The constellations are an excellent aid to quickly orient yourself in the sky, regardless of the fact that they were fanciful creations of people who had had no idea that the groups of stars they saw were nothing but coincidental projections of stars at different distances from the Earth and moving in different directions (Fig. 2). In the course of tens of thousands of years, constellations usually change and become unrecognizable. However – we hope – that will not discourage inquiring readers from their interest in the contemporary starry sky.

Where, When, and How to Find Constellations

How to begin? Preferably with the assistance of someone who can show you, directly in the sky, the most famous constellations: the Great Bear (Ursa Major), Cassiopeia, Orion, the Scorpion (Scorpius), etc. On the basis of these "starting points," with a celestial map in hand, it is relatively easy to identify other groups of stars. An excellent aid for recognizing individual constellations is the planetarium, which projects the stars and creates an artificial sky on the dome overhead. The real sky is similar: we seem to be standing in the center of an enormous hollow sphere rotating slowly around us. The stars seem to be fastened to this fictitious sphere – **the celestial sphere** – and their mutual position does not change. The apparent rotation of the celestial sphere is due to the rotation of the Earth.

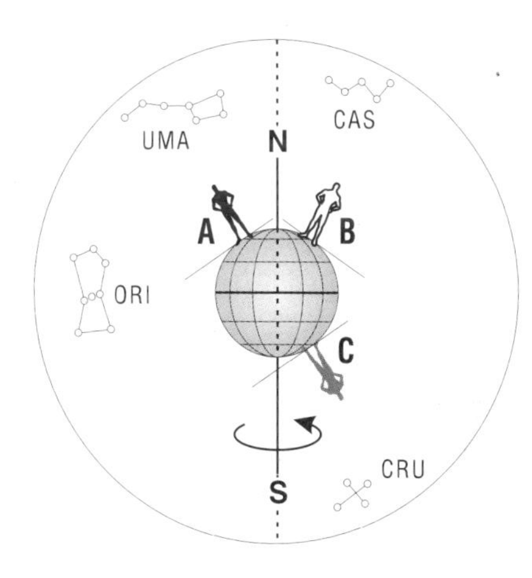

3. The celestial sphere.

In the center of the celestial sphere of infinite diameter (which we have reduced considerably in Fig. 3), a tiny ball – the Earth – rotates around its north-south axis (in Fig. 3 the Earth is magnified considerably when compared with the celestial sphere). Somewhere on the Earth, a person stands and looks at the sky. The Earth under his feet is so big that it conceals half of the celestial sphere from sight. What does the observer see in the remaining half of the celestial sphere above the horizon?

The Earth is round, as a result of which every observer sees different stars overhead (in the zenith). Mr. A lives in the northern hemisphere and can see the Great Bear (Ursa Major – UMa) overhead. Above Mr. C in the southern hemisphere, on the other hand, glitters the Southern Cross (Crux – Cru), but he can never see the Great Bear (UMa). Mr. B, with the Cassiopeia (Cas) in the zenith, lives in the same geographic latitude (on the same parallel) as Mr. A, but on the opposite meridian, so that he will see the Great Bear 12 hours later, when the Earth has rotated accordingly. From time to time, every one of the three gentlemen, A, B, and C, can see the Hunter (Orion – Ori) situated above the equator.

The selection of the constellations which we can observe successively

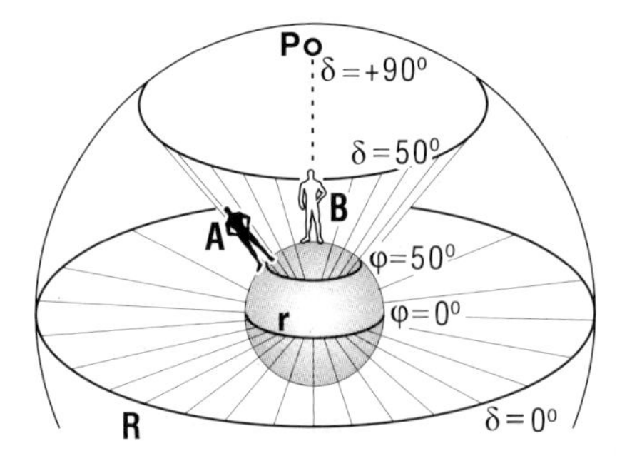

4. Latitude and declination.

on the rotating celestial sphere depends on the geographical latitude of our observation site. To be able to describe the positions of the stars on the celestial sphere more accurately, we shall project on it a grid of parallels and meridians radiating from the center of the Earth. Fig. 4 shows the projection of the Earth's equator **r** on the celestial sphere to obtain the **celestial equator R**, which divides the celestial sphere into the northern and the southern hemispheres. The picture also shows the projection of the 50th parallel of northern latitude (with Mr. A standing on it) onto the celestial sphere (δ = +50°). It is quite simple: the geographical latitude on the Earth corresponds to the **declination** in the celestial sphere. The declination on the celestial equator equals zero. North of the equator the declination is positive, south of the equator negative. Above Mr. B standing on the terrestrial north pole, there is the **celestial north pole P**, the declination of which is δ = +90°. The declination is expressed in the degrees, minutes, and seconds of arc (°, ', ") and is generally given on celestial maps.

Let us try to determine, according to Fig. 5, what part of the celestial sphere can be seen by Mr. A standing at a latitude of 50 degrees north. His **zenith Z** is intersected by the 50th parallel of northern declination (δ = +50°). That means Mr. A's zenith, which is at a declination of 50° north, forms an angle with the celestial equator of 50°. We have also projected on the celestial hemisphere Mr. A's **horizon** – the circle **H** – the horizontal plane perpendicular to Mr. A's zenith. Everything above this plane is what Mr. A can see. The zenith is always 90° above the horizon, which means that the angle from the zenith to the horizon equals exactly 90° – one fourth of the circumference of the whole circle.

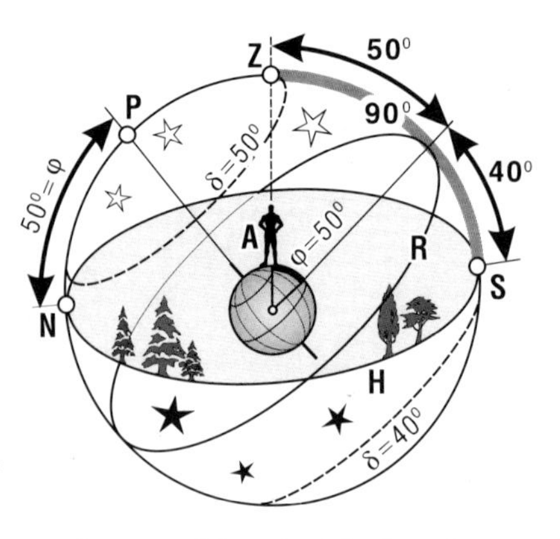

5. What part of the celestial sphere do we see?

In our case, therefore, the celes-

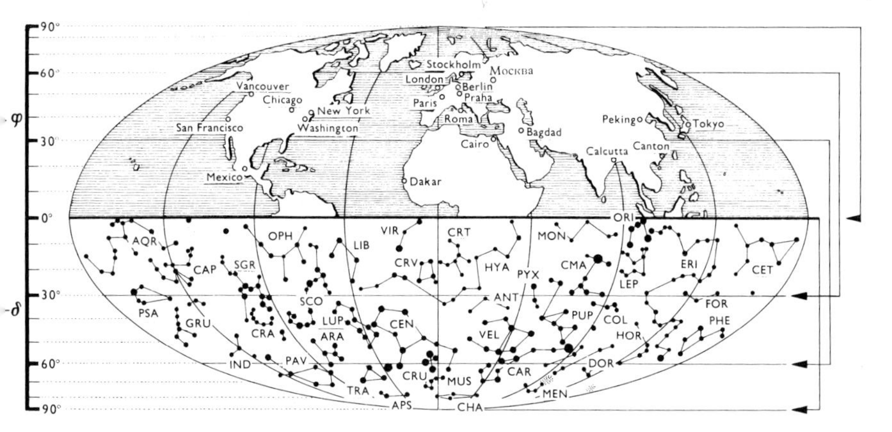

6. How far south can an observer in the northern hemisphere see? From a latitude of 90° north proceed south. This way you can determine the northernmost latitude from which you can still observe constellations in the southern hemisphere. For example, from Cairo you can see a part of the Southern Cross (Crux).

tial equator is rising above the horizon at 90° - 50° = 40°. Mr. A can see the whole eastern part of the northern hemisphere – from the celestial equator **R** to the celestial north pole **P**. Moreover, during the rotation of the Earth he can observe successively a part of the southern hemisphere as far as the 40th parallel of southern declination. And yet another important rule: the altitude, or elevation, of an object is equal to its angle above

7. Skywatchers in the southern hemisphere can see the whole southern celestial hemisphere as well as a part of the northern hemisphere as far as 90° north from their viewing latitude. From Cape Town it is possible to see the constellations of Leo (the Lion), Gemini (the Twins), Auriga (the Charioteer), and a part of the Great Bear (Ursa Major).

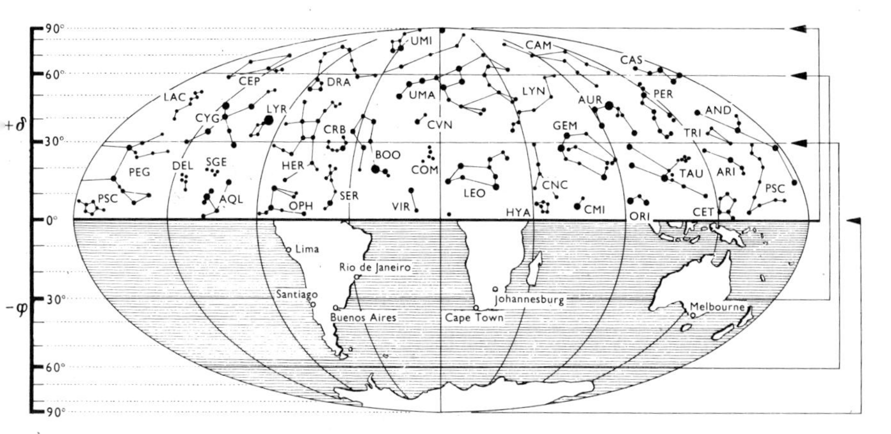

the plane of the observer's horizon. For example, the celestial north pole **P** equals the geographical latitude of the observation point (in Fig. 5, it is 50°). When stars have an angle from the pole that is smaller than φ, they are called **circumpolar stars** because they never set, or sink, below the horizon; they are always above the observer's horizon plane.

The best view of the sky is on the equator, where it is possible to see everything from the north pole right to the south pole. On the terrestrial poles, on the other hand, it is possible to see only the northern or the southern celestial hemispheres respectively. Apart from the inhospitable poles, it is impossible to see all available constellations from any given observation point simultaneously. We can only let the Earth carry us around and observe successively the rising and setting constellations apparently fastened to the celestial sphere rotating around us. Consequently, we should know or be able to foresee the turn of the celestial sphere as seen from our observation point.

First of all let us become acquainted with the **celestial meridian**. Earth's meridian is a circle that passes from pole to pole through a given location on the planet. The celestial meridian passes along the celestial sphere right above the terrestrial meridian. It is a highly important and a very useful circle in the sky passing from the north point **N** on the horizon through the celestial pole **P** and the zenith **Z** to the south point **S** – see Fig. 8. At high noon, the Sun is on the meridian because it is directly above us at our zenith. The direction of the meridian (north-south) can be determined by the compass or by means of a watch with hands: if the small hand is pointed at the Sun and the angle between the small hand and the figure 12

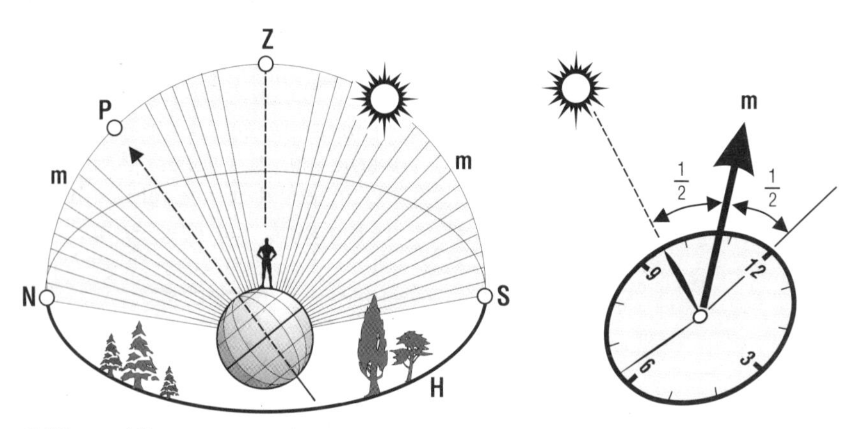

8. The meridian.

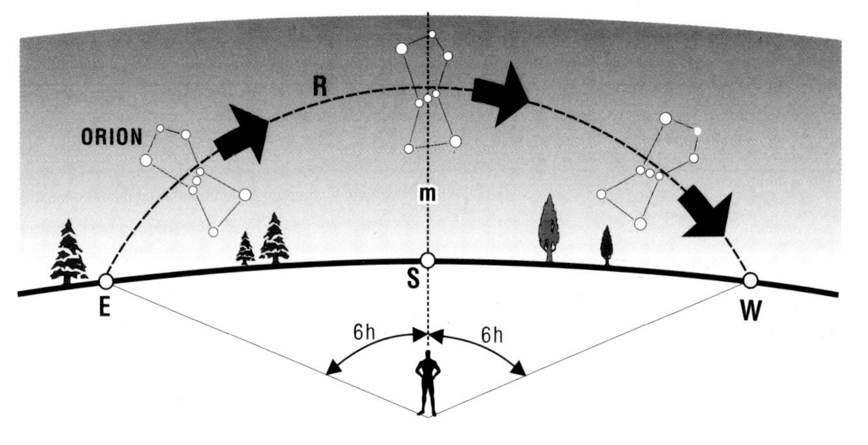

9. The stars culminate in the meridian.

is halved, the axis of this angle points straight to the meridian **m** (Fig. 8).

When in the meridian, the altitude of the Sun above the horizon is highest. Also the constellations rise highest at the meridian – we say that they culminate. Equatorial constellations, such as Orion, pass through the meridian 6 hours after their rise and 6 hours before they set (Fig. 9). With the celestial equator the meridian forms a highly accurate clock which helps us to determine how the celestial sphere is turned toward us. As the Earth rotates, our view of our celestial sphere changes; the meridian can help us know what portion of the celestial sphere we can see. We can imagine the dial of the celestial clock on the equator by dividing the equator from a certain accurately defined point – the **vernal equinox** or the **First Point of Aries** (see p. 58) – into 24 equal parts (hours) from the west to the east. Fig. 10 shows that not only the equator, but also the whole celestial sphere is divided like a peeled orange into 24 hours by the circles passing through the poles. These circles called **hour circles** are analogous to geographical meridians.

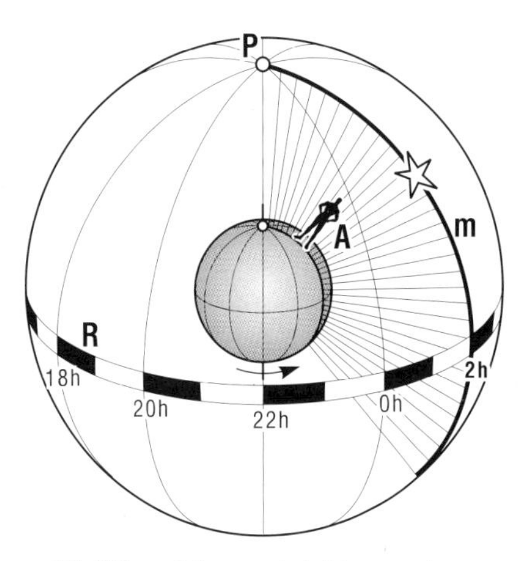

10. Sidereal time and right ascension.

Sidereal time, based on the position of the Earth to the Sun, is shown by the meridian **m**, which

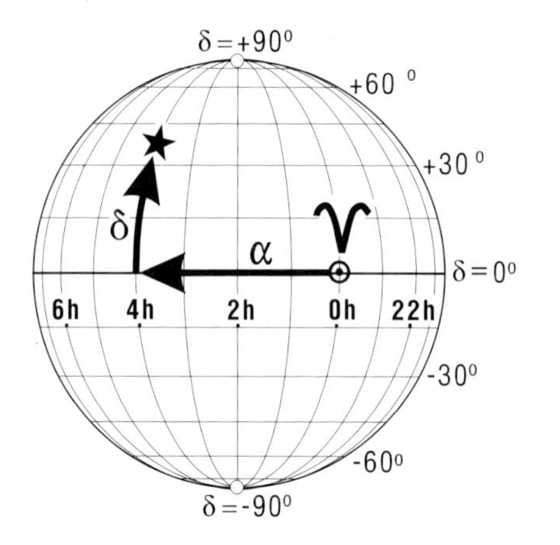

11. Equatorial coordinates of stars.

we can imagine as the edge of a giant open fan passing through our (local) geographical meridian and reaching to the sky. Like a clock hand, the meridian shows sidereal time on the equator and hour circles. Naturally, at any given moment every terrestrial meridian has a different sidereal time. On Mr. A's meridian it is 2 o'clock, sidereal time (Fig. 10). The star passing through the meridian at that moment has a **right ascension** of 2 hours. The right ascension on the celestial sphere is the same as the geographical longitude on the Earth; the declination (p. 10) is the celestial equivalent of geographical latitude.

Thus we have learned the principal system of coordinates used on celestial maps. It is the system of **equatorial coordinates** in which the position of every star is defined explicitly by two coordinates: the **right ascension** α (alpha), measured from the vernal equinox ♈ eastwards from 0 to 24 hours, and the **declination** δ (delta), measured from the equator to the poles, from 0° to 90° on either side (Fig. 11). This provides an accurate clock in the sky. Every star has its right ascension which equals sidereal time at the moment when this star passes through the meridian.

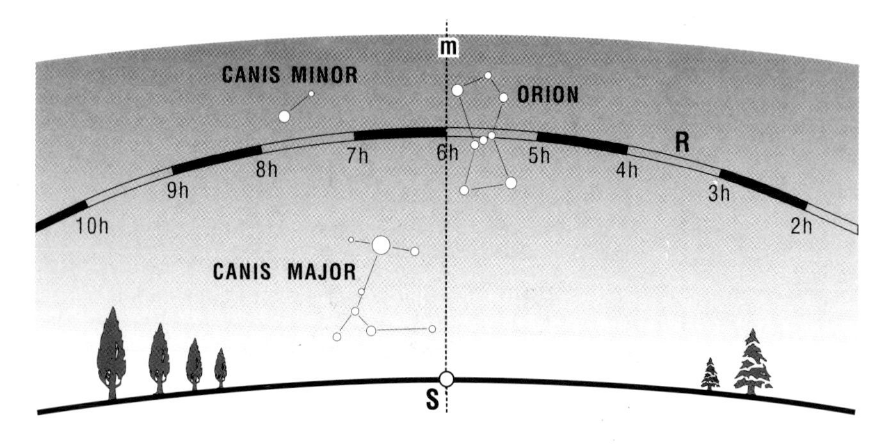

12. The visibility of constellations can be assessed by sidereal time.

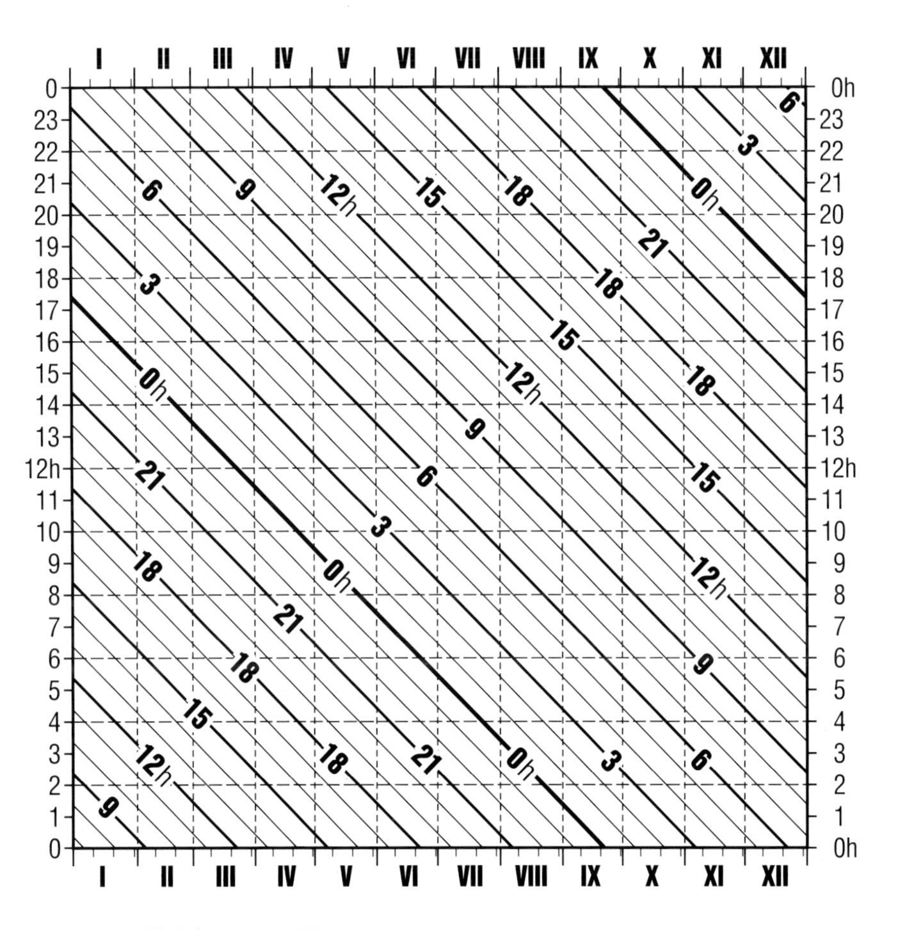

13. Diagram enabling approximate determination of sidereal time.

If we know sidereal time, we know also how the starry sky is turned to our side. At 6 o'clock, sidereal time, Orion will be always near the meridian (Fig. 12).

How can we tell sidereal time? In observatories sidereal time is shown by special clocks. However, we only need the diagram in Fig. 13. The date is read horizontally and the time vertically (this is the time of your zone, shown by your watch; but beware! during daylight savings time deduct one hour!). The sidereal time can be read on the diagonal lines. Proceed as follows: choose the observation date on the horizontal scale (e.g., May 1st) . Then choose the observation time (e.g., 20 hours, 8 p.m.). The intersection point of the horizontal and the vertical lines shows the sidereal time between two adjacent inclined lines (approx. 10:30 in this case). Sidereal time tells us which constellations are passing through the meridian. The right ascension of the stars in these constellations is close to sidereal time. All the maps in this book show right ascension in hours.

The Zodiac and the Planets

The look of the sky and the visibility of the constellations depend also on the season or, in other words, on the position of the Sun among the stars. When looking toward the Sun at the beginning of May (Fig. 14), we can see it in the direction of Aries (the Ram). Aries and the nearby constellations cannot be observed at that time, as they are above the horizon only in daytime. One month later, about June 1, we are looking from another point of the orbit of the Earth and can see the Sun in the direction of Taurus (the Bull). In this way the Sun travels in the sky from one constellation to another along a path called the **ecliptic**.

The ecliptic passes through the twelve constellations of the **zodiac**, which rank among the oldest constellations in history. Their names are: Aries, Taurus, Gemini, Cancer, Leo, Virgo, Libra, Scorpius, Sagittarius, Capricornus, Aquarius, and Pisces. These constellations originated in Mesopotamia probably as early as 5000 B.C. Initially they originated merely as the groups of stars indicating important moments of the year – equinoxes and solstices. This is obvious in the case of Libra, which symbolized the equilibrium of day and night at the time of autumn equinox some 3 or 4 thousand years ago. It may seem strange that out of the twelve

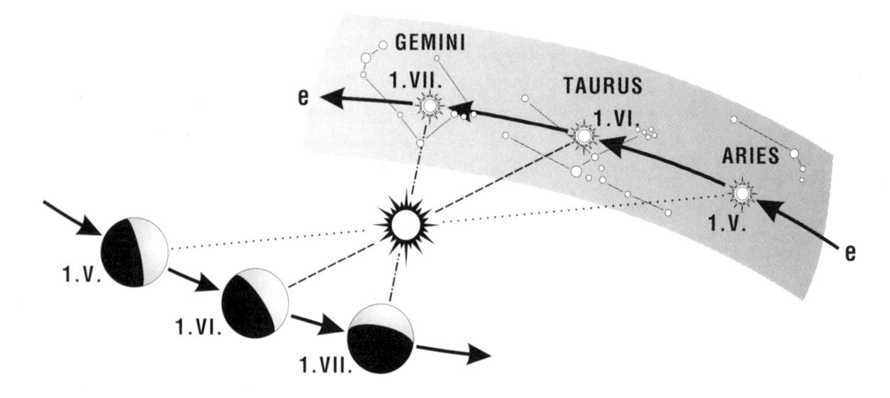

14. The constellations of the zodiac and the ecliptic.

constellations of the zodiac only seven are "zoological" in character. However, there are other inconsistencies: in actual fact the ecliptic passes through thirteen constellations! The thirteenth is the Serpent Bearer or Ophiuchus (whose part of the ecliptic is even twice as long as that of the adjacent Scorpius – see the OPH and SCO maps on pp. 159 and 185, respectively). And yet Ophiuchus does not form and has never formed part of the zodiac. The course of time and the motion of the Earth, however, left to the students of astronomy another puzzle: the **signs of the zodiac**.

♈	Aries (the Ram)	♎	Libra (the Scales)
♉	Taurus (the Bull)	♏	Scorpius (the Scorpion)
♊	Gemini (the Twins)	♐	Sagittarius (the Archer)
♋	Cancer (the Crab)	♑	Capricornus (the Goat)
♌	Leo (the Lion)	♒	Aquarius (the Water Carrier)
♍	Virgo (the Virgin)	♓	Pisces (the Fishes)

It is not only the Sun that travels through the constellations of the zodiac – so also do the Moon and planets, according to which the ancient astrologers tried to forecast the fate of men and the course of events. Around 1000 B.C., in order to better determine the positions of the planets, ancient Babylonian, and perhaps also Greek, skywatchers divided the ecliptic into 12 equal parts of 30° each ($12 \times 30° = 360°$), the 12 "signs" of the zodiac. The first sign, the first twelfth of the ecliptic, begins in the point of intersection of the ecliptic with the equator, the point in which the Sun is located on March 21. On that date the astronomical spring begins in the northern hemisphere. For this reason this point of intersection of the ecliptic with the equator is called the **vernal equinox** or the first point of Aries, because it is the beginning of the sign of Aries. Both the vernal equinox and the sign of Aries have the same symbol ♈ derived from the ram's horns. The symbols of all signs of the zodiac are shown above.

At the beginning of our era, the signs were situated in the constellations of the same name. Due to the precession of Earth's wobbling axis (see p. 58), however, the position of the vernal equinox and all signs of the zodiac slowly shift along the ecliptic. Every 26,000 years, the objects returns to their "starting point" on the ecliptic. The map on p. 59 shows that the vernal equinox and the sign of Aries are currently located in the constellation of **Pisces**. The difference between the terms **the signs of the zodiac** and **the constellations of the zodiac** is particularly prominent in the sign of Sagittarius, which interferes with three constellations at present: Scorpius, Ophiuchus and Sagittarius.

Let us leave the signs of the zodiac to historians and astrologers, and let us concentrate on the constellations. We shall learn how to recognize them in the sky. The maps on pp. 37–43 show us the position of the ecliptic. In some constellations we can often observe an "extra" bright star, not shown on the map. As a rule, it is one of the four brightest planets, Venus, Mars, Jupiter, or Saturn. The light from planets appears steady, while stars seem to twinkle. This rapid fluctuation of brightness, called **scintillation**, is caused by irregularities of the Earth's atmosphere as light travels through it.

Another difference between planets and stars is their color and brightness. The identification can be aided by an astronomical handbook and, naturally, a telescope. In most cases a good pair of binoculars is sufficient.

Venus is brilliantly white and is always the brightest planet and third-brightest object in the sky. Its brightness fluctuates between -3.5 mag and -4.5 mag (mag = magnitude, see p. 20). We see it as the morning star at sunrise or as the evening star after sunset. Its phases (crescent, first quarter, etc.) can be observed with a small telescope.

Mars is orange-red in color. Its brightness fluctuates a great deal – from +1.5 to -2.8 mag – depending on its distance from the Earth.

Jupiter shines with a white light. It is always dimmer than Venus, but brighter than Sirius, the brightest star of the sky. Its apparent magnitude varies between -1.6 and -2.3 mag. With the help of binoculars, we can see its four brightest moons.

Saturn is faintly yellowish, its magnitude varying from +0.9 to -0.1 mag. Its well-known rings can be observed with a telescope.

From time to time we can see another planet, Mercury, even without binoculars. It can be seen at twilight, either shortly before sunrise or not long after sunset. Information on the visibility of Mercury is given in astronomical yearbooks, where we can find also the data required for the observation of other planets (with telescopes only!) – Uranus, Neptune and Pluto.

The planets travel near the ecliptic and their paths among the stars recall complex loops. This is because we are observing the planets orbiting around the Sun from the also-moving Earth. Other members of the solar system, planetoids and comets, follow orbits that often deviate considerably from the ecliptic. Therefore, these objects can be observed not only in the constellations of the zodiac, but also in other constellations.

A CELESTIAL GUIDE

DEEP SPACE OBJECTS	20 – 31
CELESTIAL MAPS	32 – 45
CONSTELLATION MAPS	46 – 215

DEEP SPACE OBJECTS – 1
Stars

Stars are very hot bodies of plasma (mostly hydrogen gas) radiating their own light. They have much greater masses than the planets, and radiate large amounts of energy created by thermonuclear reactions that take place within them. Mutual comparisons of the diameters, masses, luminosity and other characteristics of stars are often made relative to the Sun (1 sun, 1 $\odot$), which is an average star.

The stars can be seen in the sky as points of different brightness. The scale of the star brightness is their **apparent magnitude** (mag). The faintest stars visible with the naked eye are approximately of the sixth magnitude (6.0 mag). Very bright stars are of 1.0 mag (Spica), brighter stars are of zero magnitude (0.0 mag, e.g., Vega), and the brightest stars are of negative magnitudes (Sirius is of -1.4 mag).

The distances of the stars from the Earth are vastly different, which, naturally, influences their magnitudes. To compare the actual **luminosity of stars**, we convert their apparent magnitude to the so-called **absolute magnitude M**, the brightness stars would have if they were all at a uniform distance of 32.6 light-years from the Earth. One **light-year** is the distance light travels in one year: 9.46×10^{12} km (about 9.5 trillion km).

The distance of nearby stars can be determined on the basis of their parallaxes (see below). Parallax is the angular displacement in the apparent position of a star when observed from two widely separated points. The farther the star, the smaller the angle, which is formed by observing the star from two different points on Earth's orbit **o** around the Sun **S**, appears. However, this trigonometric method fails at distances exceeding hundreds and thousands of light-years, and the distances of deep space bodies are determined by much less accurate methods. Therefore, the reader should not be surprised if the data on the distance of the same object, given in different works, differ considerably.

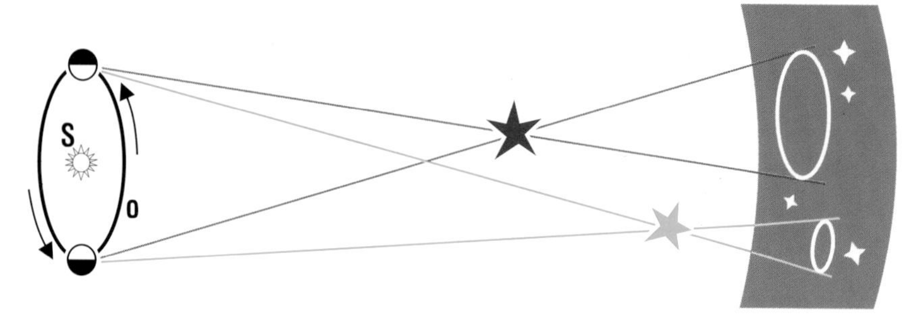

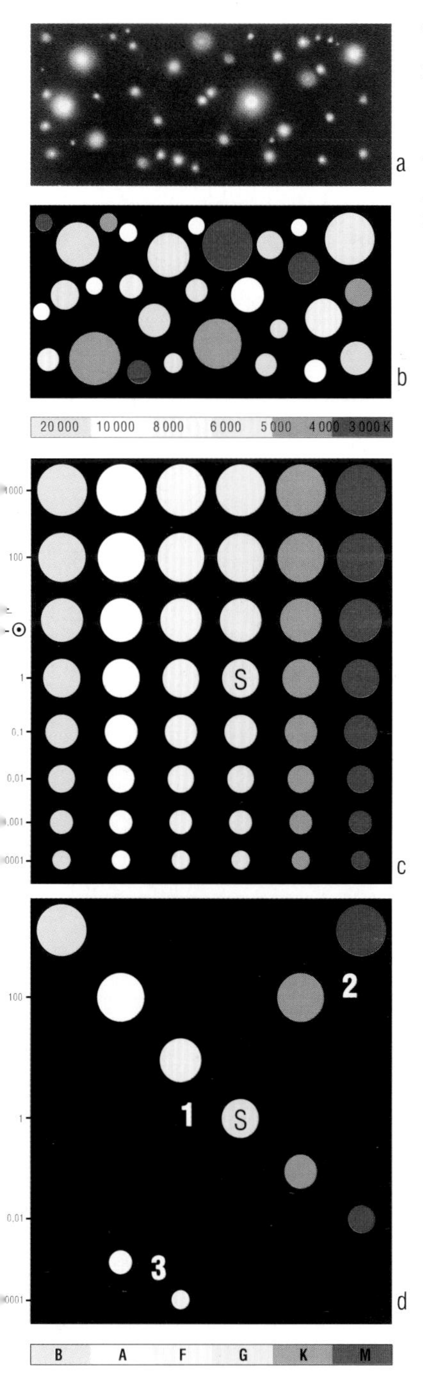

The luminosity and temperature of individual stars differ. The hottest stars with surface temperatures above 10,000 K (Kelvin) shine mostly with blue light, the coldest stars with temperatures of about 3,000 K are red (**a**). We have depicted the luminosity of stars by discs of different size (**b**). The greater the luminosity, the larger the disc. A star with a larger surface area has a greater luminosity than a smaller star of equal temperature. The colors correspond to the temperatures of the stars according to the scale below.

Let us classify now the chaotic medley of stars, and arrange them in a diagram according to their size (luminosity **L** in relation to the luminosity of the Sun $\mathbf{L}_{\odot}$) and according to color (temperature) by grouping discs of the same color in columns. The Sun's luminosity and surface temperature (6,000 K) will serve as units of measurement. You would then think that with the enormous number of stars around us, we would find every combination of sizes (luminosity) and color (temperature), as shown in (**c**). But you'd be surprised.

Even if we take into account the greatest number of stars, the discs form a patterned diagram, leaving other parts void (**d**). That is because of the evolution of stars and one of its laws: The hotter the star, the brighter it is. Thus we have arrived, although in considerably simplified form, at an important **temperature-luminosity diagram**, known also as the Hertzsprung-Russel Diagram (**HRD**).

Giants and Dwarfs

In the temperature-luminosity diagram (**HRD**; p. 21**d**) the absolute majority of stars including the Sun **S** range in the **Main Sequence (1)** running across the diagram as a continuous band from the upper left to the lower right. Other important star types are **stellar giants** and **supergiants (2)** and **white dwarfs (3)**. They are really fitting names. A giant must be a real giant for its luminosity – a red giant, for example – to be a million times as high as that of a small star of the same temperature.

A great deal of information about the temperature and other characteristics of a star can be derived from its **spectrum**, formed by breaking up the star's light into individual color components with a glass prism or a diffraction grating. The spectrum of each star corresponds to its temperature, and the stars are ranked into **spectral classes** accordingly. Differences among them are not due to their different chemical composition, but due primarily to their different surface temperatures. All stars consist mostly of hydrogen and helium; other elements occur in much lower quantities. The picture below gives examples of the spectra of some bright stars of different temperature. The wavelengths are given in nanometers (nm).

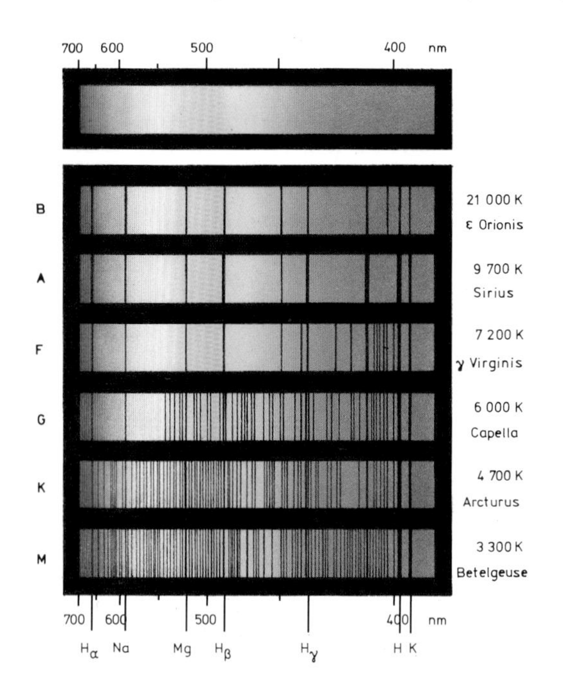

On the continuous colored background, the stellar spectra contain dark (absorption) spectral lines – the "signatures" of chemical elements. According to the occurrence and thickness of certain spectral lines, the stars are divided into spectral classes defined by the letters **O, B, A, F, G, K, M** (O marks the hottest stars). Every class is divided into 10 subclasses, such as G0, G1 through G9; the latter is very close to subclass K0.

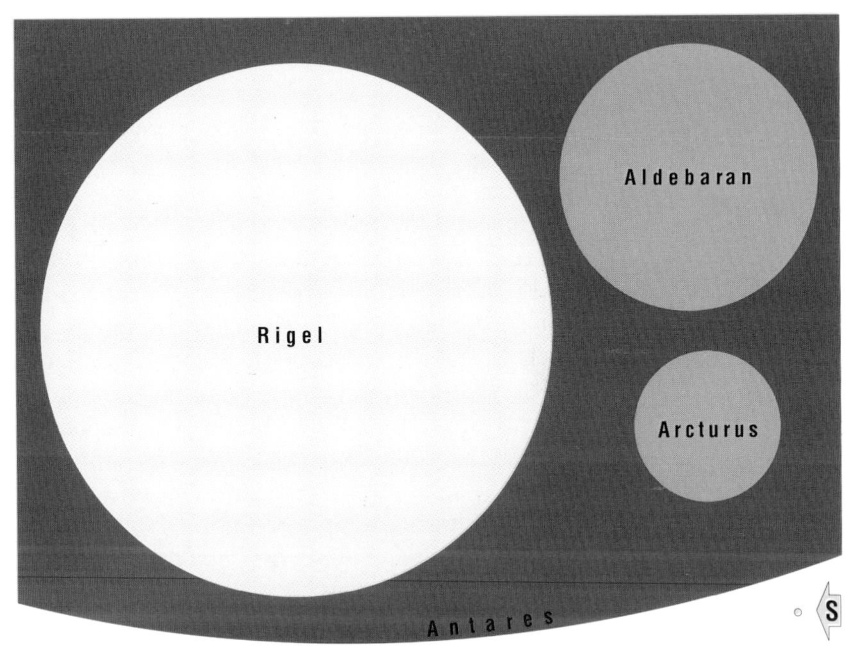

The picture above shows the relative sizes of several giants and super-giants compared with the Sun **S**, the disc of which has a diameter of only 1 millimeter.

The picture below shows a part of the Sun **S** (the diameter of which would be 110 cm here) in comparison with some red and white dwarfs and the Earth **E**.

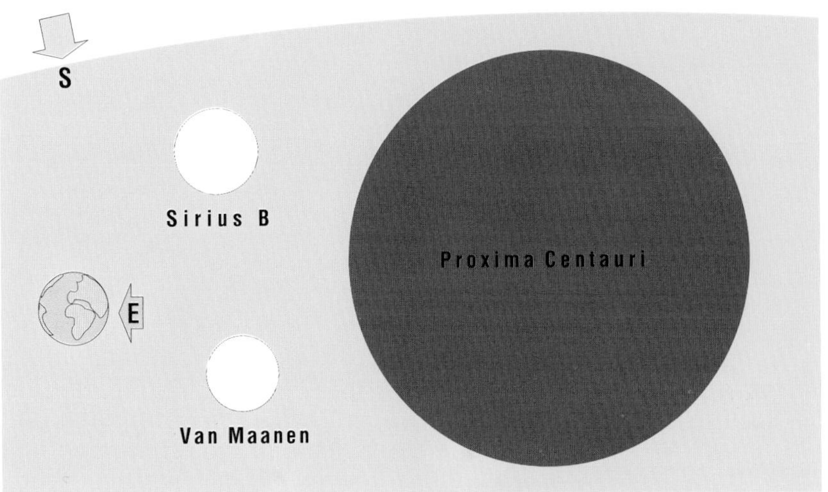

DEEP SPACE OBJECTS – 3
Double Stars and Variable Stars

Double stars and multiple stars include more than half of all stars we see in the sky. The constellation maps show the best-known examples. If the components of a double star can be separated with a telescope, we speak about **visual double star**, which may be **physical** (two stars that orbit one another) or **optical** (unrelated stars with different distances from the Earth that appear along the same line of sight from our perspective). Visual double stars rank among the most popular objects of amateur observation and serve as an excellent test of the resolving power of a telescope. If the components are so close to each other that they cannot be resolved visually and their binary character is revealed only in the spectrum, we speak about a **spectroscopic binary**. Especially interesting are **eclipsing binaries**. The component stars eclipse one another, which manifests itself by periodic variations of brightness. A typical example is Algol in Perseus.

Variable stars vary their brightness, regularly or irregularly, for physical reasons – by, for example, changes in their diameter and surface temperature. Our maps show some characteristic examples of their many types. Most common among them are pulsating variable stars, called **cepheids**. The "signature" of every variable star is its **light curve**, the graphic representation of the changes of its brightness plotted against time.

Extraordinary changes of brightness are due to explosive processes taking place in the final evolutionary stages of the stars. Due to their sudden increase of brightness, they are classified as **novae** ("new stars") and **supernovae**, if their sudden outburst is extraordinarily powerful.

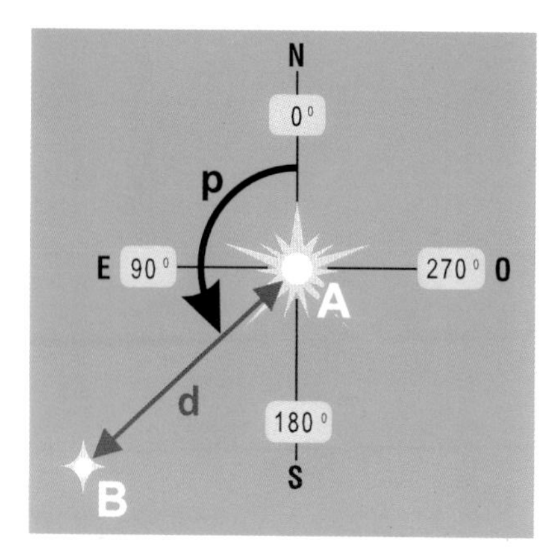

There are two fundamental parameters characterizing double stars. Their **separation d** is the angular distance of the double star components, usually specified in arc seconds. The **position angle p** gives the direction from the brighter component to the fainter component, measured in angular degrees from the north **N** (0°) to the east **E** (90°), the south **S** (180°) and the west **W** (270°).

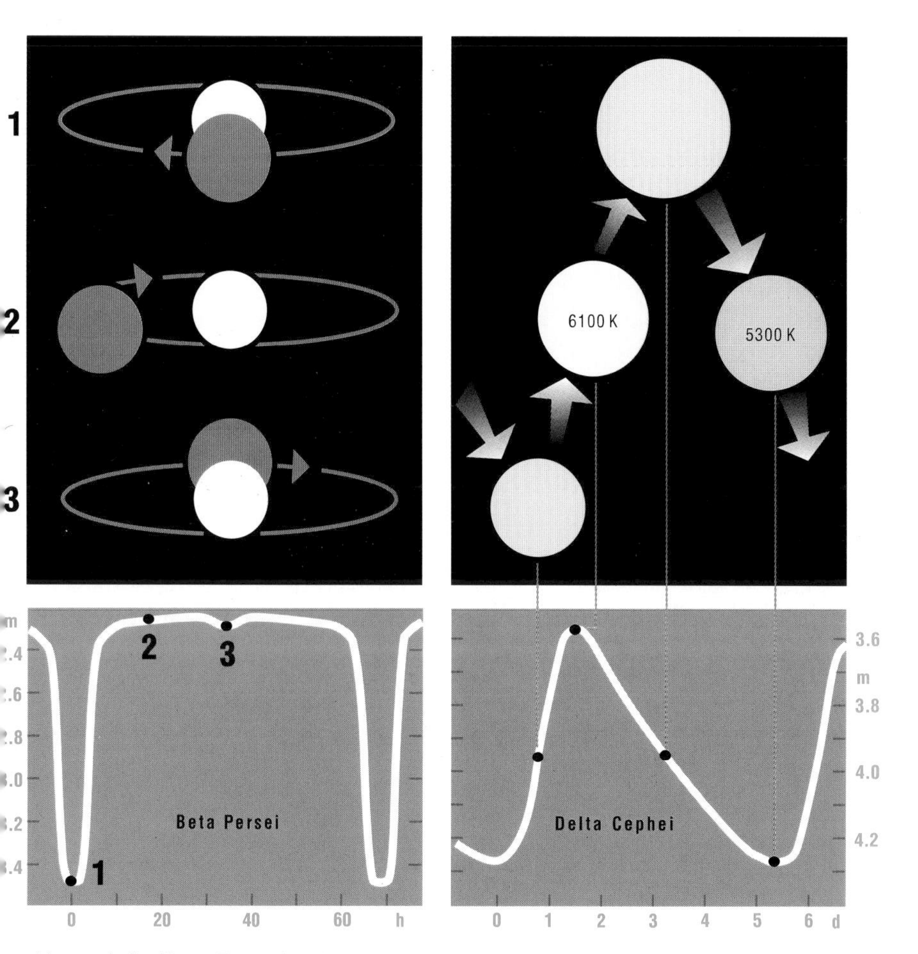

Above left: **Beta Persei – Algol** is a typical **eclipsing binary**. This fainter, cooler orange star with a surface temperature of 4,500 K and a diameter 3.5 times the diameter of the Sun eclipses at regular intervals the brighter and hotter star with a surface temperature of 13,000 K and a diameter of 2.9 times the diameter of the Sun. The brightness of Algol, therefore, varies between the maximum of 2.25 mag to the minimum of 3.5 mag with a period of 68.8 hours. The variation is shown by the light curve below (lower left).
Above right: **Delta Cephei** is an example of a pulsating variable star – a **cepheid**. It is a yellow giant with a diameter of 45 suns. The star pulsates, alternately inflating and deflating. The temperature fluctuates from 5,300 K to 6,100 K. In its maximum the star is 2,900 times, in its minimum 1,300 times as bright as the Sun. The magnitude of Delta Cephei varies from 3.5 to 4.4 in a period of 5.366341 days. Its light curve is shown below (lower right).

DEEP SPACE OBJECTS – 4

The Milky Way – The Galaxy

The Sun and all the stars we see in the sky form only a minute part of an enormous stellar island called the **Milky Way Galaxy**. The distance from the Sun to the center of the Milky Way, situated in the direction of Sagittarius, is less than 30,000 light-years. The Milky Way is a spherical formation, but most of its stars are concentrated in a flat, spiral disc. The central lens-shaped part of the disc has a diameter of some 100,000 light-years and contains several hundreds of billions of stars. The clouds of interstellar dust and gas cumulate along the plane of the Galaxy. If we could look at the Milky Way from the side at a distance, these clouds would appear as a dark separating band, which is shown schematically in the picture on the opposite page.

Here and there we can observe the interstellar matter of bright or dark **nebulae**. Near the plane of the galaxy, there are **open star clusters**. The disc-shaped plane is surrounded with a spherical "halo" with a diameter of 100,000 light-years containing old stars and **globular star clusters**.

If we look around from the Sun **S** in the plane of the galactic disc, the multitude of stars around us and the spiral arms of the Galaxy project onto the celestial sphere as a silvery hazy band – the Milky Way. Aside from the galactic plane there are "galactic windows," vistas of the world of distant galaxies.

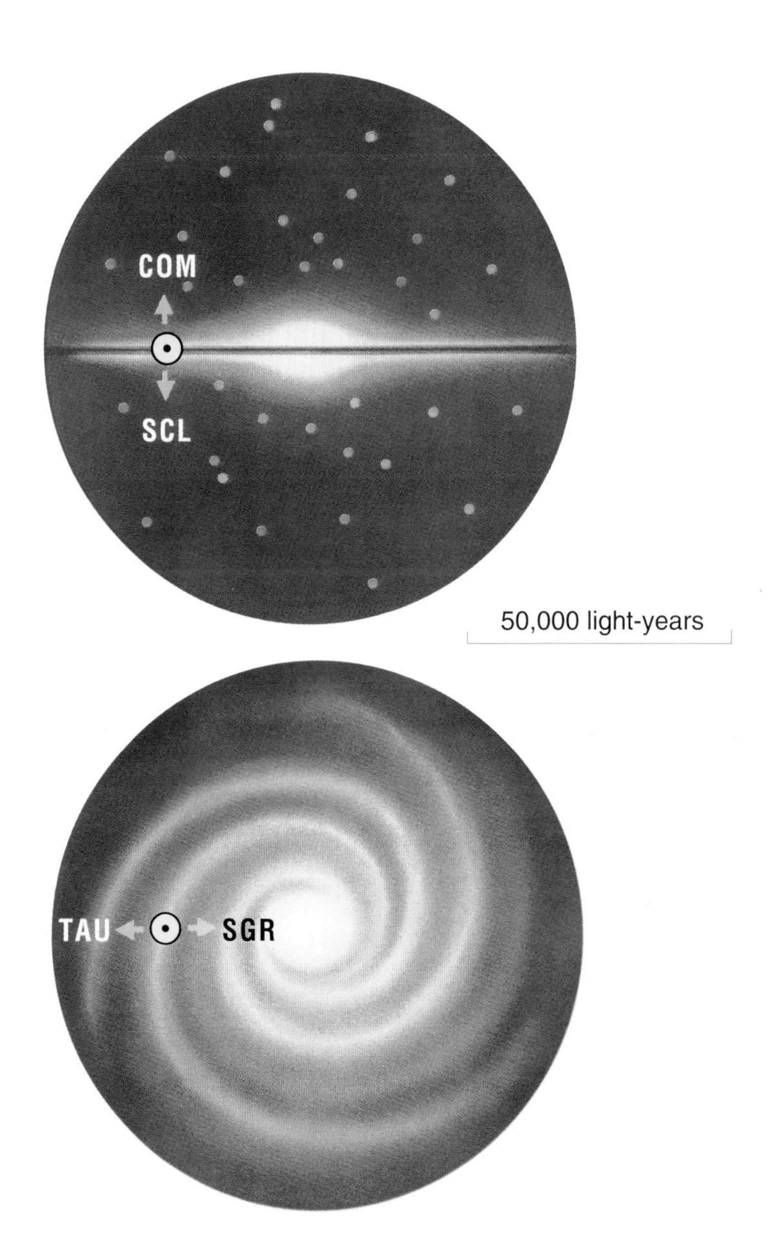

Top: a schematic side view of our Galaxy. Below: a view of the Galaxy from above. The yellow disc shows the position of the Sun.

DEEP SPACE OBJECTS – 5
Clusters and Nebulae

Clusters and nebulae are among the most favorite objects of amateur observers. Some of them are visible as hazy clouds with fuzzy contours with the naked eye, binoculars, or a small telescope. A selection can be seen in our constellation maps. However, only major telescopes can show their whole beauty. New observation technologies make it possible to record their faint radiation, which is invisible to the naked eye. Deep space objects are fascinating not only for what we see, but primarily for the "invisible" clues we have about them thanks to the research of modern astronomy.

Open clusters consist mostly of young, hot stars which have originated from a common nebula, undergone common development and are mutually linked by gravitation. They have dozens to hundreds of components and their actual diameter varies between 5 and 50 light-years. We know more than one thousand of them. They occur in the proximity of the Milky Way plane.

Globular clusters contain tens of thousands to millions of stars, which are among the oldest in the Galaxy. The shape of globular clusters is spheroid, with the density of stars increasing toward their center. The actual diameter of globular clusters varies between 50 and 300 light-years. We know about 150 of them in our Galaxy.

Nebulae are clouds of interstellar gas and dust. In the proximity of very hot stars their radiation causes the gas (mostly hydrogen) clouds to emit characteristic red light and to manifest themselves as **diffuse emission nebulae** or H-II regions. The prevailing blue hue of **reflection nebulae**, on the other hand, is due to the scattering of the light of cooler stars on the particles of interstellar dust.

Planetary nebulae originate around certain aging stars that have ejected the outer layers of their stellar atmosphere. The central stars of these nebulae are extraordinarily hot (30,000–150,000 K) and, therefore, initiate the radiation of nebular gases. Numerous planetary nebulae appear to the observer as small disc-like planets – hence the name.

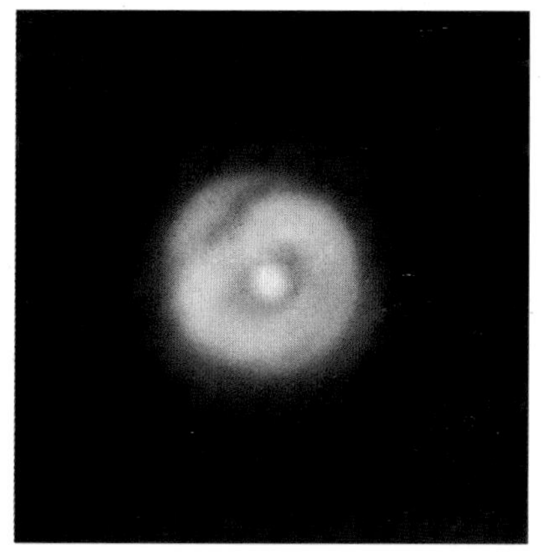

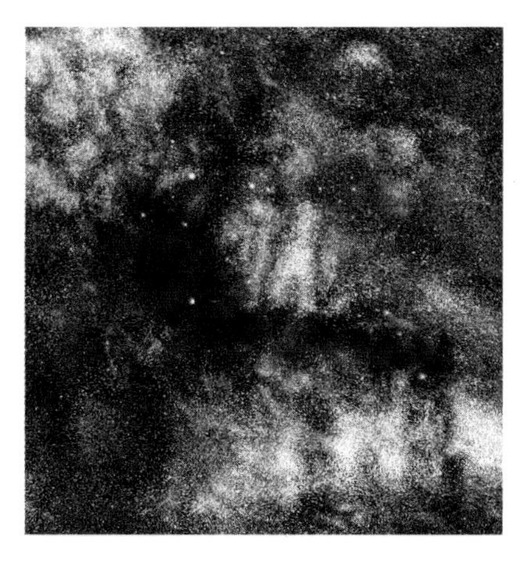

Dark nebulae, or **absorption nebulae**, are dark clouds of interstellar dust and gas distant from bright stars and, therefore, not luminous. However, they are distinct against the background of the Milky Way. In combination with bright nebulae they often form fantastic patterns. All nebulae are the material for the formation of further stars.

Galaxies

Galaxies are the basic building blocks of the universe. They are stellar systems containing hundreds of millions to hundreds of billions of stars and, naturally, considerable quantities of interstellar matter. Some resemble our Galaxy, others differ from it both in mass and in structure. Minor telescopes enable us to see only the brightest and nearest galaxies. Most of them appear only as indistinct hazy clouds, but big telescopes make it possible to photograph even individual stars, clusters and nebulae in the nearest galaxies. In color photos, their spiral arms appear bluish thanks to the radiation of young hot stars originating in them.

Galaxies congregate into groups or systems of various size. Our Galaxy forms part of the so-called **Local Group**, which contains some 30 to 40 galaxies, including the well-known M 31 galaxy in Andromeda (p. 46).

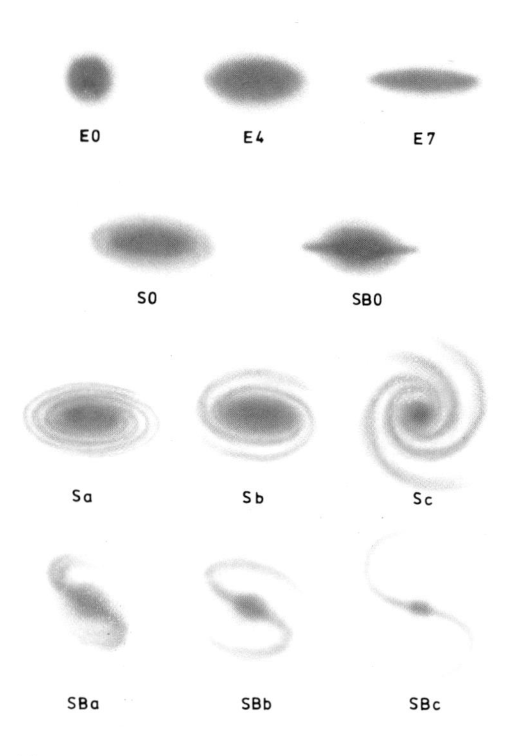

Hubble's Classification of Galaxies. According to their appearance the galaxies are classified as **elliptical** (E), **spiral** (S) or **barred spiral** (SB). Spiral galaxies are divided into subclasses a, b, and c according to their development of spirals and the relative size of their nucleus. The spiral arms of SB galaxies link up with a "bar" radiating from the nucleus. Galaxies of irregular shape are designated as Irr.

In space the galaxies are oriented in different directions (with reference to us, to the Earth). Therefore, the view of some of them facilitates our imagining what our galaxy would look like when observed from that distance.
The picture above shows the M 101 Galaxy in the constellation Ursa Major observed "from above." The picture below shows the NGC 891 galaxy in Andromeda, which we observe "from the side."

Big clusters of galaxies contain hundreds to thousands of galaxies; giant clusters, up to hundreds of thousands of galaxies. The observation of galaxies has resulted in the discovery of the expansion of the universe. It has come to light that the distances among the galaxies are continuously increasing, the velocity of expansion growing with the distance of the galaxies. The velocity of expansion manifests itself in the spectrum of the galaxy by the shift of spectral lines toward the red end of the spectrum (the so-called redshift). This phenomenon is used for the determination of the distance of the galaxies from the Earth.

This section of this book contains two sets of star maps: general maps of the whole sky on pp. 33–45 and the maps of individual constellations on pp. 47–215.

General maps are intended for initial orientation in the sky and for identification of the individual constellations. The ancient constellation figures have disappeared. And only their names, the complicated boundaries between them, and the names and positions of the brightest stars within the constellation have remained on modern star maps. The bright stars can be connected by straight lines into easy-to-remember figures.

Constellation maps acquaint the reader with deep space objects, their positions, names etc. All maps are printed on a uniform scale. To give the reader an idea of the constellation size in the sky, every map contains the seven star asterism of the Big Dipper for scale. The abbreviation of the individual constellations (see the list on p. 216) are differentiated by means of three colors: red for northern constellations, orange for the constellations intersected by the celestial equator, and yellow for southern constellations. Bright stars are represented with their traditional names or small letters of Greek alphabet (so-called Bayer's letters), accompanied by the genitive of the Latin name of the constellation or its three-letter abbreviation, Beta Leonis or Beta Leo, for example. Also star designation by the so-called Flamsteed numbers, like 70 Ophiuchi, is used frequently. Bright star clusters, nebulae and galaxies (the so-called deep-sky objects) are best known under the number of the Messier's catalogue or under their proper names, such as M 20 – Trifid (p. 183). Otherwise these objects are often designated by the catalogue numbers of the NGC (New General Catalogue) or the IC (Index Catalogue). Our maps comprise stars up to 5.2 mag and a selection of interesting deep space objects (including some of the fainter ones), many of which are available even for amateur observation.

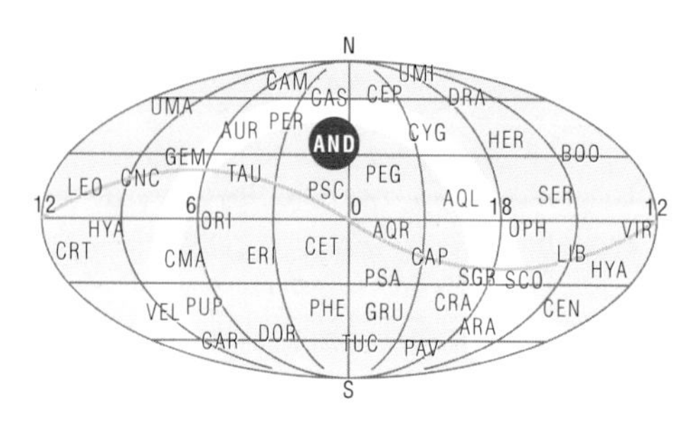

Orientation maps show the position of the respective constellation (shown by a color disc – in this particular case Andromeda) on the celestial sphere. The ecliptic is yellow, the band of the Milky Way is shown schematically in white.

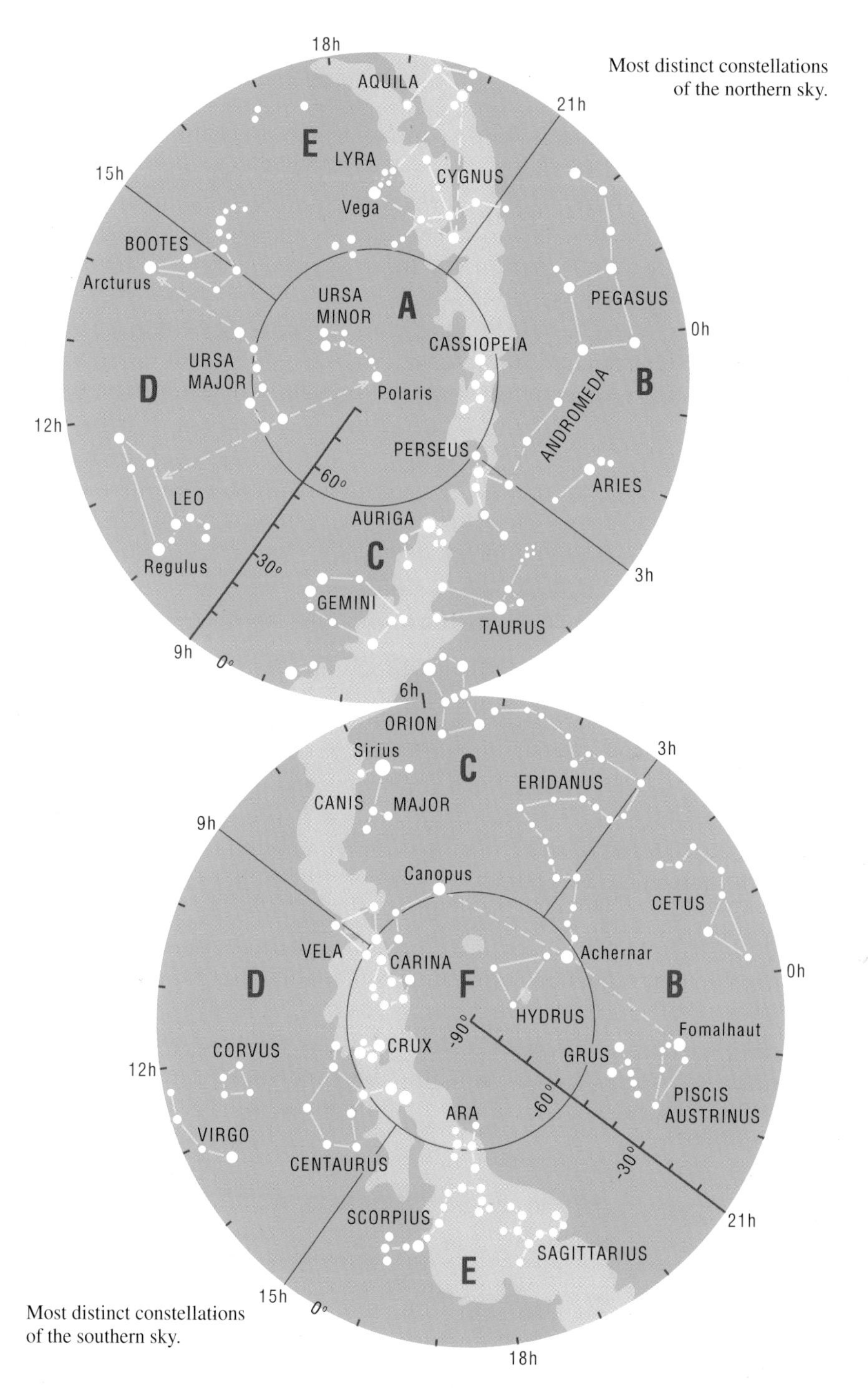

Most distinct constellations of the northern sky.

Most distinct constellations of the southern sky.

Constellations in the Region of the North Celestial Pole

The opposite page presents the first of six maps showing the whole northern and southern skies. They show the general layout and the boundaries of the individual constellations, mapped in greater detail on pp. 47–215. Equatorial coordinates, right ascension and declination, are shown on all celestial maps in this atlas and are valid for the epoch of 2000.0 (see pp. 58 and 59 on precession).

The most distinct constellation in the north pole region is **Ursa Major** (the Great Bear), including the characteristic asterism of seven stars known as the Big Dipper. In the northern hemisphere, it is the best-known constellation, and everyone who begins to get acquainted with the sky should start with it. The line connecting the stars Dubhe and Merak aims at the North Star (Polaris) in the constellation of **Ursa Minor** (the Little Bear). The North Star is located less than one degree from the North Pole (center of the map). The constellation of **Draco** (the Dragon) with a quadrangular head winds between the two Bears.

If you proceed from the Great Bear across the North Star to the opposite side of the pole, you can find there, at the same distance from the pole, the five brightest stars of **Cassiopeia** in the form of W or inverted M, another well-known constellation of the northern sky. In its proximity you can find the double open star clusters Chi and h Persei, visible as a hazy spot with the naked eye. Between Cassiopeia and the Dragon, the indistinct constellation of **Cepheus** can be observed. The location of **Camelopardalis** (the Giraffe) and **Lynx** is characterized by amateur astronomers that "they are where you cannot see a thing," as they consist of faint stars visible only when the local observing conditions are very good.

The map shows also the 50°th parallel of northern declination. All constellations within this circle appear as circumpolar, never-setting, to the observer at 40° northern geographical latitude. Generally speaking, circumpolar stars have the angular distance measured from the north pole smaller than the geographical latitude of the observation site (as shown in Fig. 5, p. 10).

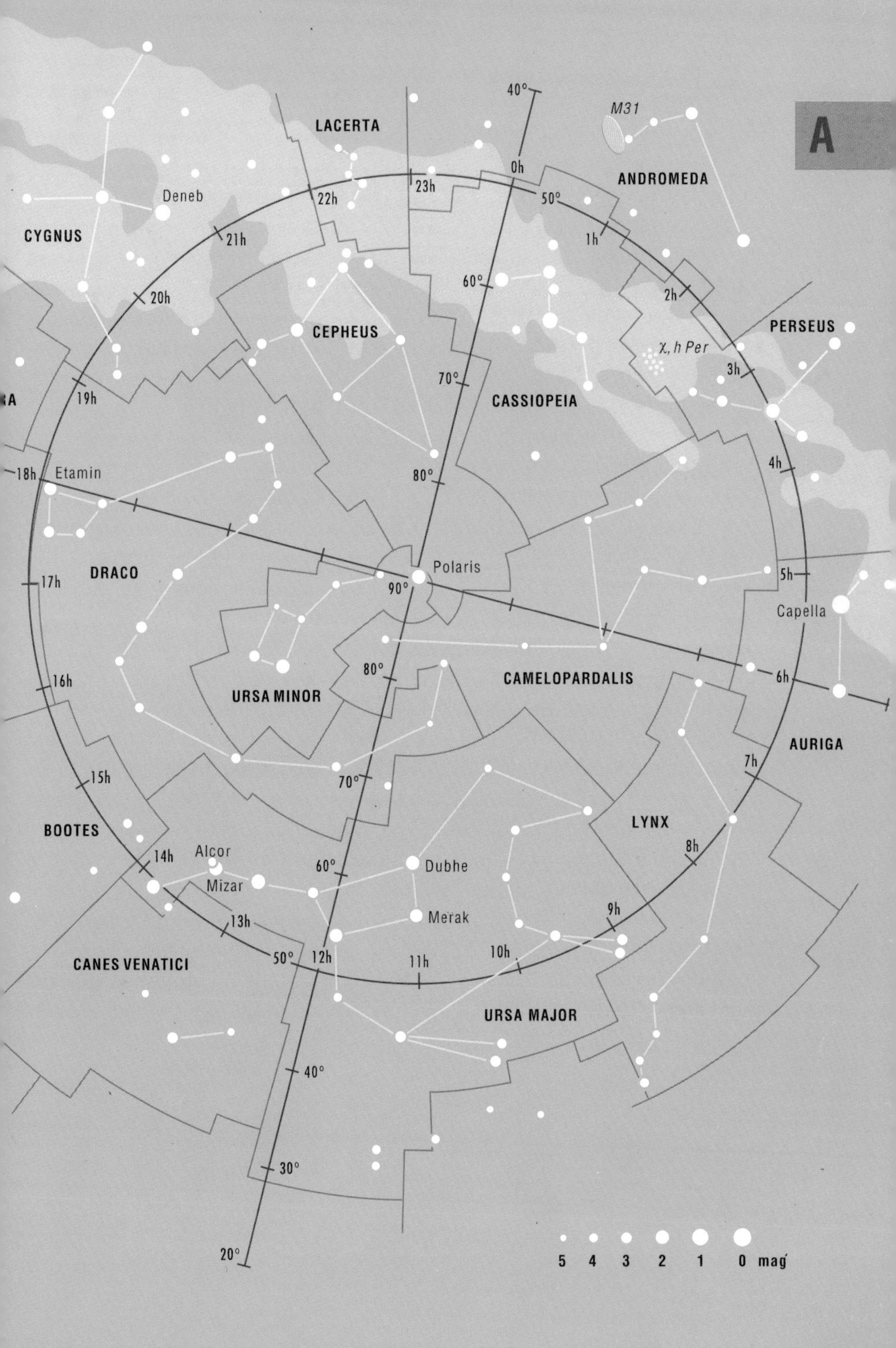

A
LACERTA
M31
ANDROMEDA
CYGNUS
Deneb
CEPHEUS
PERSEUS
χ, h Per
CASSIOPEIA
Etamin
DRACO
Polaris
URSA MINOR
CAMELOPARDALIS
Capella
AURIGA
LYNX
BOOTES
Alcor
Mizar
Dubhe
Merak
CANES VENATICI
URSA MAJOR
5 4 3 2 1 0 mag

Autumn Constellations

The maps on pp. 37–43 divide the constellations into four groups according to the seasons when they can be observed for the better part of a night in the northern hemisphere.

The autumn sky is dominated by the so-called **Great Square** defined by Alpha, Beta and Gamma **Pegasi** and the Sirrah star from the adjacent **Andromeda**. If we also add to this "square" the stars Mirach and Alamak from Andromeda and Algol from Perseus, we obtain a figure conspicuously similar to the magnified asterism of the Big Dipper. That is the principal orientation figure which will form the basis of our observations. Above it, there is the characteristic **W** of **Cassiopeia**, below Andromeda there are two minor constellations called **Triangulum** (the Triangle) and **Aries** (the Ram).

The southeast corner of the Great Square aims at the open "V" of **Pisces** (the Fishes). It is rather inconspicuous, like its neighbor **Aquarius** (the Water Carrier), which can be remembered best by the characteristic group of stars on the equator, known from pictorial sky maps as "the jug" from which the water carrier pours water. Somewhat more distinct is **Capricornus** (the Goat) situated on the ecliptic to the southwest from the triangular "head" of Pegasus. The letter "V" of Pisces aims, like a big arrow, at the center of the constellation of **Cetus** (the Whale).

In the southern hemisphere you can find primarily the brightest stars: Fomalhaut from the constellation of **Piscis Austrinus** (the Southern Fish) and Achernar from the elongated **Eridanus** (the River Eridanus). Together with Canopus from **Carina** (the Keel), the above mentioned stars form a conspicuous threesome situated approximately on the same line, which can well serve our orientation (see the map on p. 33). South of Fomalhaut is **Grus** (the Crane); in the proximity of Achernar is **Phoenix** (the Phoenix). Faint stars of **Fornax** (the Furnace), **Sculptor** (the Sculptor) and **Microscopium** (the Microscope) can be observed only if conditions are very good and only from the sites where these parts of the sky rise sufficiently high above the horizon.

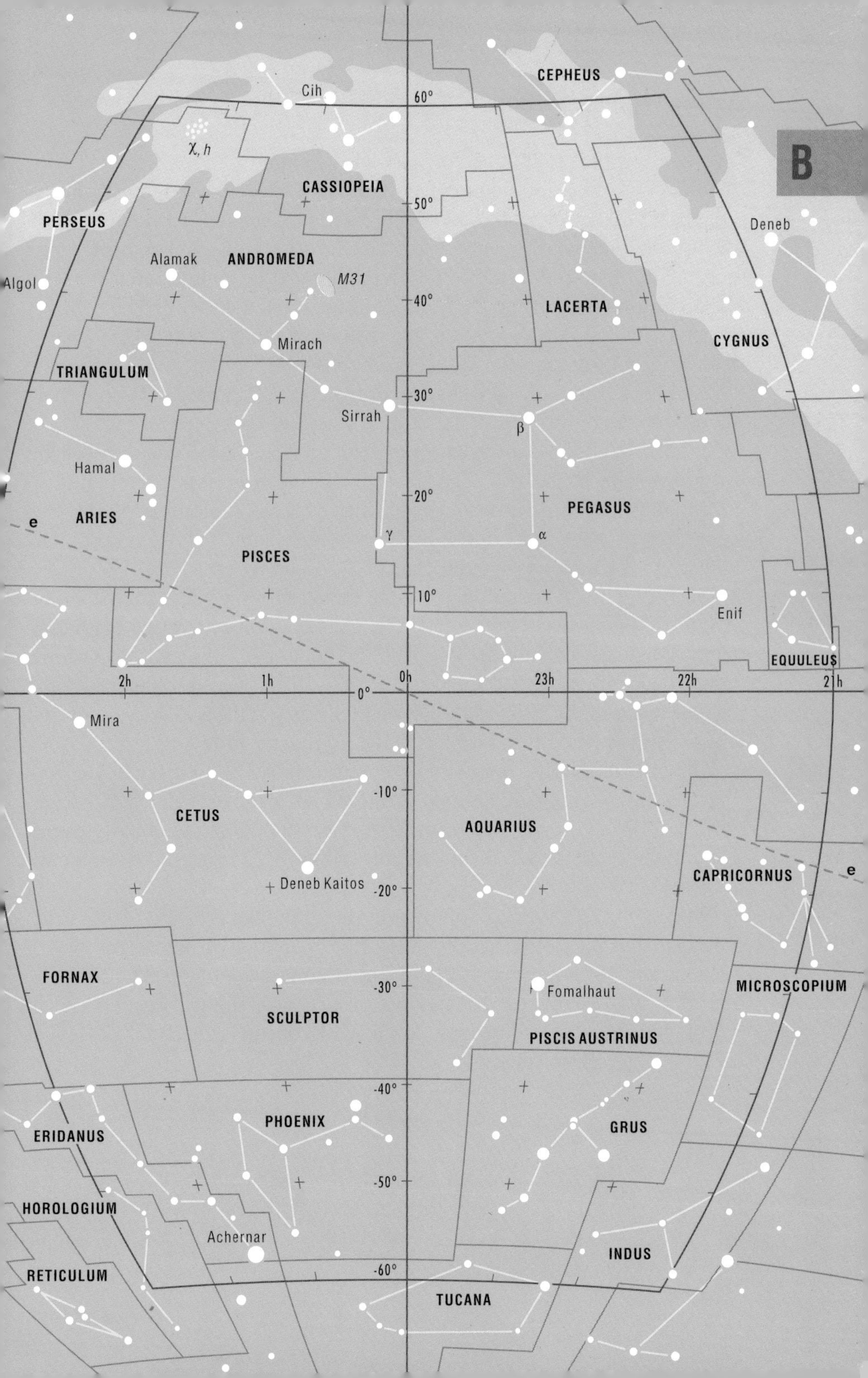

B
CEPHEUS
Cih
60°
χ, h
CASSIOPEIA
50°
PERSEUS
Deneb
Alamak
ANDROMEDA
M31
Algol
40°
LACERTA
Mirach
CYGNUS
TRIANGULUM
30°
Sirrah
β
Hamal
20°
ARIES
PEGASUS
e
γ
α
PISCES
10°
Enif
EQUULEUS
2h
1h
0h
23h
22h
21h
0°
Mira
-10°
CETUS
AQUARIUS
CAPRICORNUS
e
Deneb Kaitos
-20°
FORNAX
-30°
Fomalhaut
MICROSCOPIUM
SCULPTOR
PISCIS AUSTRINUS
-40°
PHOENIX
ERIDANUS
GRUS
-50°
HOROLOGIUM
Achernar
INDUS
-60°
RETICULUM
TUCANA

Winter Constellations

The winter sky is rich in bright stars and orientation in it is easy even for the beginner. The dominant constellation is **Orion**, the brightest stars of which form a marked quadrangle with three stars regularly arranged across the middle. The whole group resembles the open wings of a butterfly. According to the initial figure of Orion the hunter, the central three stars are called Orion's belt. Its northernmost star is situated very near the equator. When extended northwards the line of Orion's belt points to the orange Aldebaran in **Taurus** (the Bull). Further along this line you can find the conspicuous group of the Pleiades. Between the Pleiades and Cassiopeia there is a bridge formed by the arch of **Perseus** stars. Not far away there is Capella from the constellation of **Auriga** (the Charioteer), one of the brightest stars in the sky. The line connecting Rigel and Betelgeuse in Orion aims at **Gemini** (the Twins) which you can remember by the pair of its stars, Castor and Pollux. In their vicinity is the inconspicuous constellation of **Cancer** (the Crab).

According to an ancient legend, the hunter Orion was accompanied by two dogs, one big and one small. The constellation of **Canis Major** (the Great Dog) can be found along the line connecting the stars of Orion's belt. The brightest star of the Great Dog – Sirius – is simultaneously the brightest star of the whole sky. Between Sirius and the Twins you can find **Canis Minor** (the Little Dog) with the bright star Procyon.

The orientation in winter sky is served well by the "**winter hexagon**," the vertices of which are defined by the stars Capella – Aldebaran – Rigel – Sirius – Procyon – Pollux with Castor. Less popular, but equally useful for orientation purposes, is the "**winter triangle**" of Procyon – Betelgeuse – Sirius.

Proceeding from Sirius southwards along the band of the Milky Way, you can find the constellations **Vela** (the Sails) and **Carina** (the Keel). Canopus or Alpha Carinae is the second brightest star of the sky (after Sirius) and was used often for navigation of space flights. From Rigel in Orion you can observe the meandering of **Eridanus** (the River Eridanus) as far as the bright Achernar.

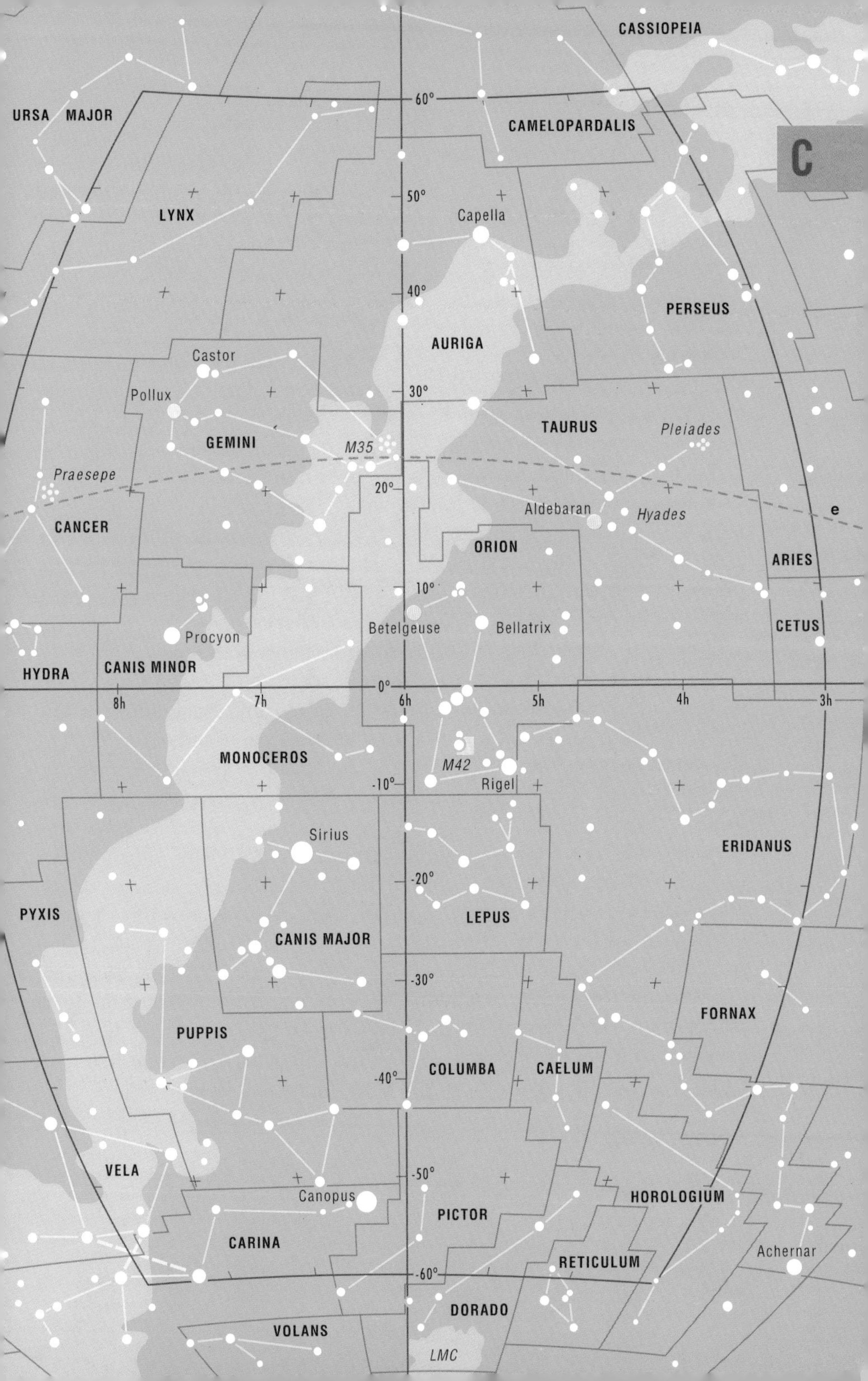
C
CASSIOPEIA
URSA MAJOR
CAMELOPARDALIS
LYNX
Capella
AURIGA
PERSEUS
Castor
Pollux
GEMINI
TAURUS
Pleiades
M35
Praesepe
CANCER
Aldebaran
Hyades
e
ORION
ARIES
Betelgeuse
Bellatrix
CETUS
Procyon
HYDRA
CANIS MINOR
8h
7h
6h
5h
4h
3h
MONOCEROS
M42
Rigel
Sirius
ERIDANUS
PYXIS
LEPUS
CANIS MAJOR
FORNAX
PUPPIS
COLUMBA
CAEELUM
VELA
Canopus
PICTOR
HOROLOGIUM
CARINA
RETICULUM
Achernar
DORADO
VOLANS
LMC
60°
50°
40°
30°
20°
10°
0°
-10°
-20°
-30°
-40°
-50°
-60°

Spring Constellations

There are great differences between the northern and the southern hemispheres during spring. Near the celestial equator and further northward, bright stars are scarce; the south offers the view of one of the richest parts of the sky. The paucity of the northern part of the sky is apparent. But it is in these very voids, aside from the Milky Way, that "galactic windows" are situated offering the vistas into the most distant depths of the universe with alien galaxies and galactic clusters.

As the starting point in the northern sky we shall use the Big Dipper in the constellation of **Ursa Major** (the Great Bear). The line connecting the stars Dubhe and Merak aims southwards at the constellation of **Leo** (the Lion), which can be recognized easily by the group of stars recalling a big sickle or an inverted question mark. According to the initial representation, these stars depicted the lion's mane. The Lion's brightest star, Regulus, is situated on the ecliptic. If you extend the handle of the Big Dipper (the "tail" of the Great Bear) by an arch to the equator, you find the orange Arcturus in the constellation of **Boötes** (the Herdsman). Proceeding along this arch below the equator, you find the bright Spica in the constellation of **Virgo** (the Virgin). The three stars, Regulus – Arcturus – Spica, form the so-called **spring triangle**, which will serve us well in orientation. Below the triangle there is the elongated constellation of **Hydra** (the Water Snake), the "head" of which can be found below Cancer. Between Hydra and the Virgin there is the small, but beautiful constellation of **Corvus** (the Crow) and the less distinct **Crater** (the Cup) next to it.

South of Hydra there are beautiful parts of the Milky Way that contain the two best-known southern constellations – **Crux** (the Southern Cross) and **Centaurus** (the Centaur). You don't have to travel to the southern hemisphere to see them. The Southern Cross is visible low above the horizon from the Tropic of Cancer (23.5° north latitude). Without travelling far, you can also see the Southern Cross on the artificial sky of the local planetarium. Be careful not to mistake it for the "false cross" at the boundary between **Vela** (the Sails) and **Carina** (the Keel)! The real Southern Cross is in the vicinity of the bright stars Alpha and Beta Centauri.

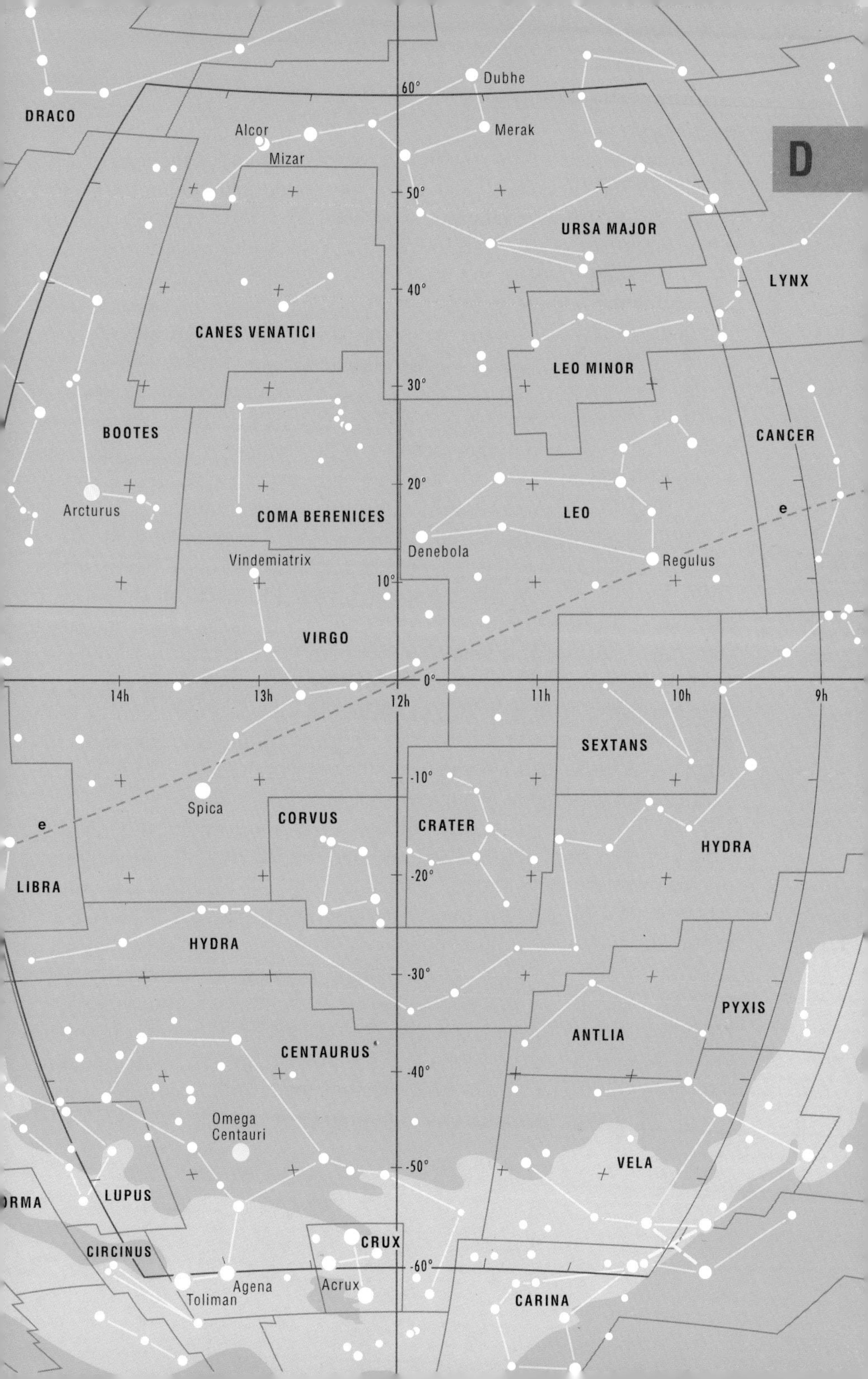
D
DRACO
Dubhe
Merak
Alcor
Mizar
URSA MAJOR
LYNX
CANES VENATICI
LEO MINOR
BOOTES
CANCER
Arcturus
COMA BERENICES
LEO
Denebola
Regulus
Vindemiatrix
VIRGO
e
14h
13h
12h
11h
10h
9h
60°
50°
40°
30°
20°
10°
0°
-10°
-20°
-30°
-40°
-50°
-60°
SEXTANS
Spica
CORVUS
CRATER
HYDRA
e
LIBRA
HYDRA
PYXIS
ANTLIA
CENTAURUS
Omega
Centauri
VELA
LUPUS
CRUX
CIRCINUS
Agena
Acrux
Toliman
CARINA

Summer Constellations

In favorable observation conditions, the silvery band of the Milky Way is a reliable guide to constellations. The most important orientation figure is the so-called **summer triangle**, whose vertices consist of the bright stars Vega, Deneb and Altair. The bluish white Vega in the constellation of **Lyra** (the Lyre) ranks among the brightest stars of the summer sky. Deneb is situated in the tail of **Cygnus** (the Swan) flying with spread wings south along the Milky Way. In the Swan, the Milky Way divides into two arms. Altair in the constellation of **Aquila** (the Eagle) is accompanied by a fainter star on either side, remotely recalling the belt of Orion. The summer triangle can also be observed in the first half of the night during autumn. The night sky is darker then than in the summer, revealing the beauty of the Milky Way. North of Altair there are two of the smallest constellations: **Sagitta** (the Arrow) and **Delphinus** (the Dolphin).

North of Vega you can find the head of the Dragon and west of Lyre you can look for the extensive, but not very distinct constellation of **Hercules**, the faint stars of which form an inverted K. Between Hercules and the Herdsman (see "Spring Constellations") there is the small, but easy to remember **Corona Borealis** (the Northern Crown) with the bright star Gemma. Below the Crown is **Serpens Caput** (the Serpent's Head). The long Serpent's body is held by **Ophiuchus** (the Serpent Bearer), to the northeast of which you can observe **Serpens Cauda** (the Serpent's Tail) reaching as far as the Eagle between the arms of the Milky Way. The Serpent is a unique case of a "binary" constellation.

From the ecliptic (zodiacal) constellations in the brightest parts of the Milky Way, **Scorpius** (the Scorpion) and **Sagittarius** (the Archer) stand out the most. On the Scorpion's chest, it is possible to observe the red star Antares, west of which is the constellation of **Libra** (the Scales). South of the Scorpion you can find the distinct constellation of **Ara** (the Altar) recalling the shape of a simple armchair. Beyond the Scorpion you can proceed along the Milky Way to the constellation of **Lupus** (the Wolf). From the "teapot" formed by the brightest stars of the Archer you can proceed eastwards to the constellation of **Capricornus** (the Goat). South of the Archer there is a regular bow of faint stars forming the small constellation of **Corona Australis** (the Southern Crown).

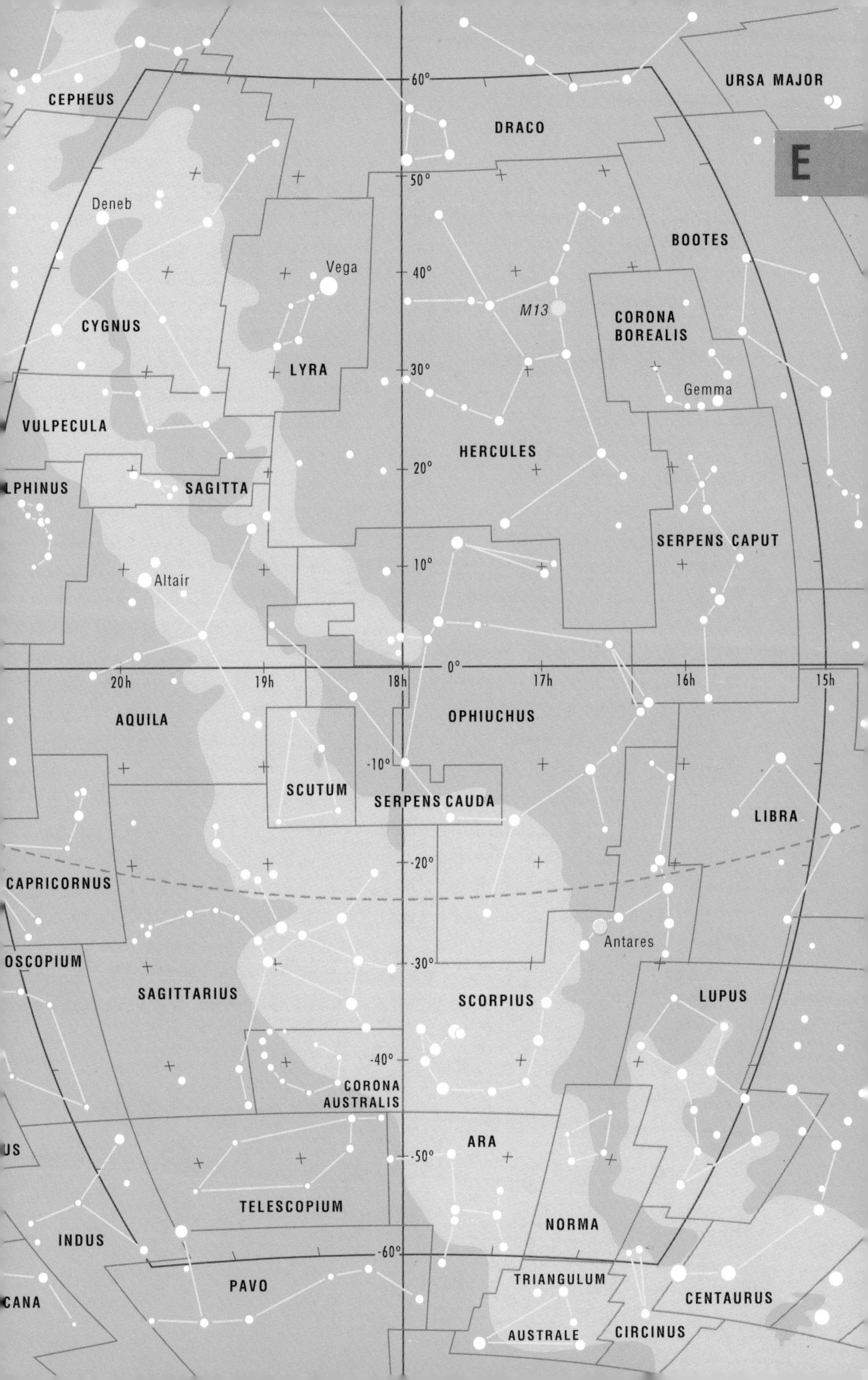
E
CEPHEUS
DRACO
URSA MAJOR
Deneb
Vega
CYGNUS
LYRA
M13
BOOTES
CORONA BOREALIS
Gemma
VULPECULA
HERCULES
LPHINUS
SAGITTA
SERPENS CAPUT
Altair
20h
19h
18h
0°
17h
16h
15h
AQUILA
OPHIUCHUS
SCUTUM
SERPENS CAUDA
LIBRA
CAPRICORNUS
Antares
OSCOPIUM
SAGITTARIUS
SCORPIUS
LUPUS
CORONA AUSTRALIS
ARA
US
TELESCOPIUM
INDUS
NORMA
PAVO
TRIANGULUM
CENTAURUS
CANA
AUSTRALE
CIRCINUS
60°
50°
40°
30°
20°
10°
-10°
-20°
-30°
-40°
-50°
-60°

Constellations in the Region of the South Celestial Pole

To the inhabitants of the northern hemisphere this area is exotic, indeed. However, it too has a sufficient number of starting points. You can find your way most easily in the band of the Milky Way where the constellation of **Crux** (the Southern Cross) provides a reliable orientation point. Next to the Southern Cross, there seems to be a dark hole in the Milky Way – actually a dark nebula called the Coal Sack. The Pointers, the line connecting the bright stars Alpha and Beta Centauri, point at the Southern Cross. These pointers assist us in the differentiation of the Southern Cross from the "false cross" at the border of the constellations **Vela** (the Sails) and **Carina** (the Keel). Among the number of minor constellations we can find **Musca** (the Fly) near the Southern Cross and **Triangulum Australe** (the Southern Triangle) near Alpha Centauri.

There is no bright star in the proximity of the celestial south pole like the North Star in the northern hemisphere. The south pole is situated in the highly indistinct constellation of **Octans** (the Octant), which does not help us much in orientation. The position of the south pole can be estimated roughly by means of the Southern Cross – the longer arm of which points to the pole – or by means of Magellanic Clouds forming with the pole an approximately equilateral triangle. Both Magellanic Clouds (Small Magellanic Cloud – SMC, and Large Magellanic Cloud – LMC) are visible even in moonlight. They are minor galaxies in the immediate vicinity of our Galaxy.

A handy orientation axis in the southern sky is the line connecting the very bright stars Canopus in the Keel, Archenar in Eridanus and Fomalhaut in the Southern Fish. Along this line you can try to find the minor and less distinct constellations: **Pictor** (the Painter), **Dorado** (the Swordfish), **Reticulum** (the Reticle), **Horologium** (the Pendulum Clock) and **Phoenix** (the Phoenix). Among other constellations near the south pole you can find relatively easily **Pavo** (the Peacock). The point of guidance will be its solitary bright star Alpha Pavonis, also called the Peacock.

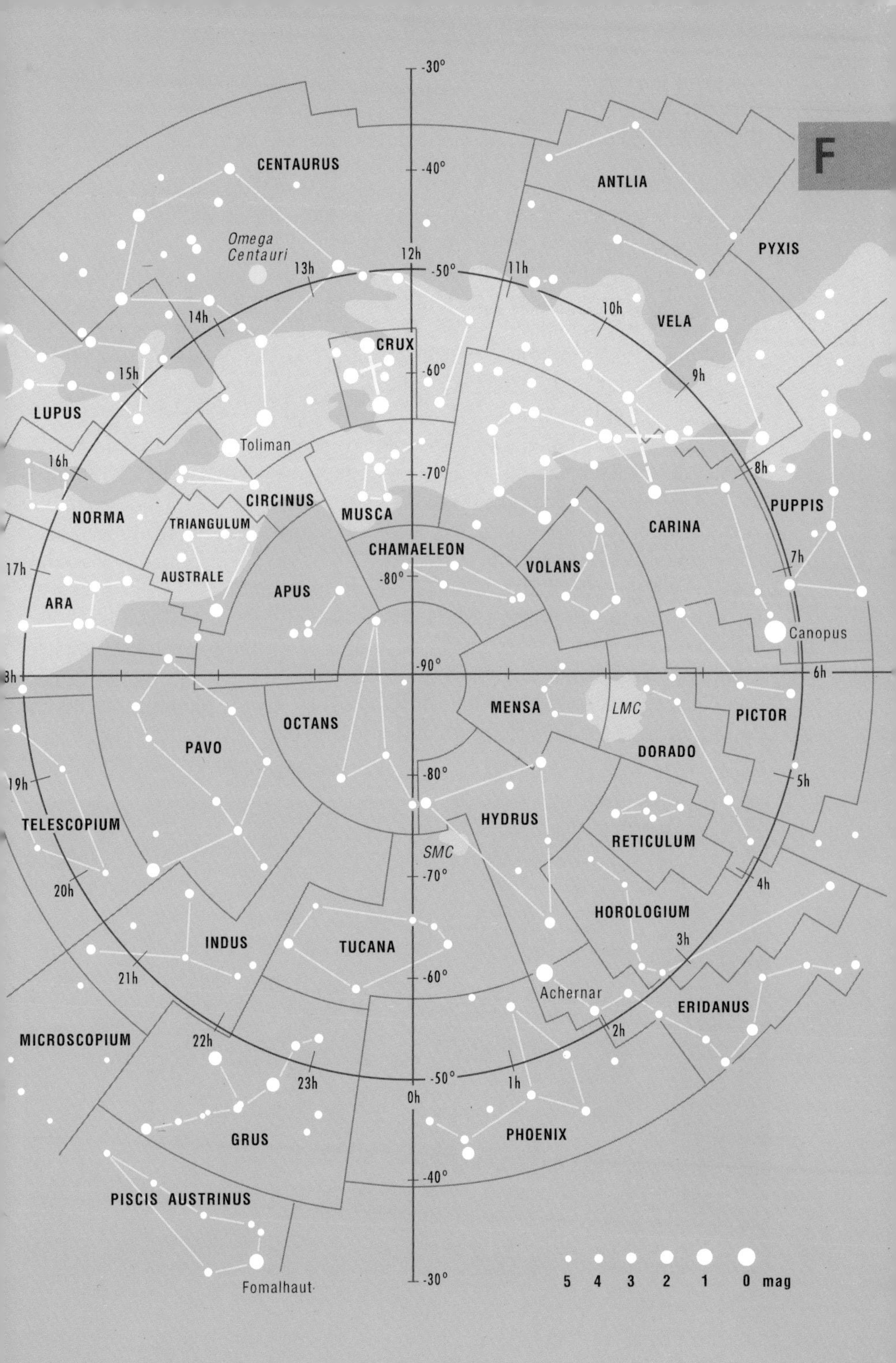

F
CENTAURUS
ANTLIA
PYXIS
Omega Centauri
VELA
CRUX
LUPUS
Toliman
CARINA
PUPPIS
NORMA
CIRCINUS
MUSCA
TRIANGULUM AUSTRALE
CHAMAELEON
VOLANS
ARA
APUS
Canopus
MENSA
LMC
PICTOR
OCTANS
DORADO
PAVO
HYDRUS
TELESCOPIUM
RETICULUM
SMC
HOROLOGIUM
INDUS
TUCANA
Achernar
ERIDANUS
MICROSCOPIUM
GRUS
PHOENIX
PISCIS AUSTRINUS
Fomalhaut
5 4 3 2 1 0 mag

ANDROMEDA

The old constellation maps usually show Andromeda chained to the rock as a sacrifice to the Whale (Cetus). It was the punishment of the gods for the offense committed by her mother, queen Cassiopeia, who boasted that she was more beautiful than the Sea Nymphs, the daughters of Neptune. You can find Andromeda near the adjacent "Great Square." Its brightest stars are **Alpha – Sirrah**, **Beta – Mirach**, and **Gamma – Alamak**. The visual magnitude of all three varies about 2.1 mag and their distance from the Sun is 97, 200 and 355 light-years respectively. Beta And is a red giant of a diameter equal 30 suns.

Galaxy M 31 – NGC 224 (the Great Andromeda Galaxy, formerly the Great Andromeda Nebula) is the nearest big spiral galaxy, the member of the Local Group of galaxies. Its distance from the Earth is 2.9 million light-years, its diameter approximately 170,000 light-years. There are four satellite galaxies in its proximity, viz., **M 32 – NGC 221**, situated 0.4° to the south, **M 110 – NGC 205**, situated 0.6° to NW, and **NGC 185** and **NGC 147**, situated in Cassiopeia. M 31 can be seen by the naked eye as an elongated misty cloud, which has a diameter of 3° and over in binoculars. Small telescopes can discern a distinctly brighter center; major instruments distinguish numerous details. However, only the photographs taken by big telescopes show the stars and other objects within the galaxy.

Galaxy **M 31** with satellites **M 32** (right) and **M 110** (upper left).

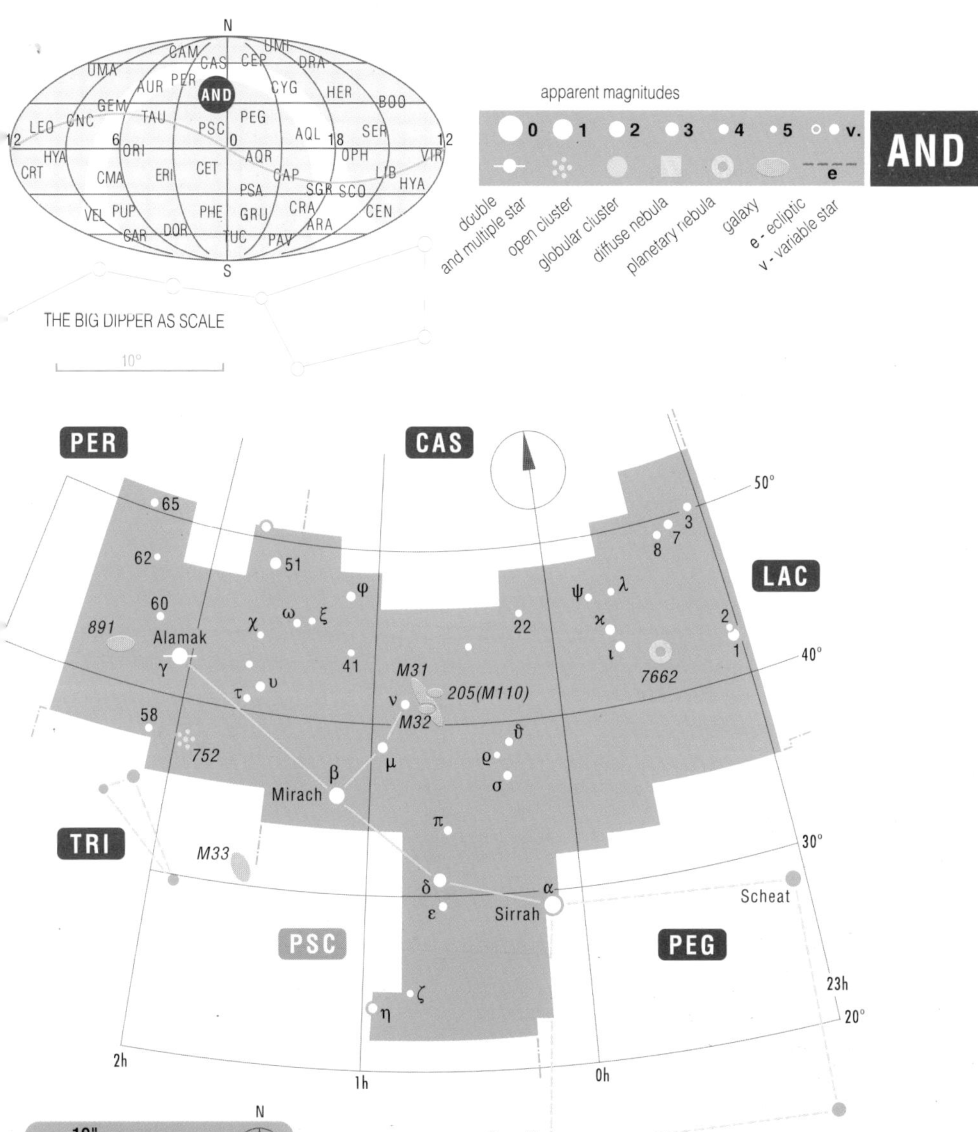

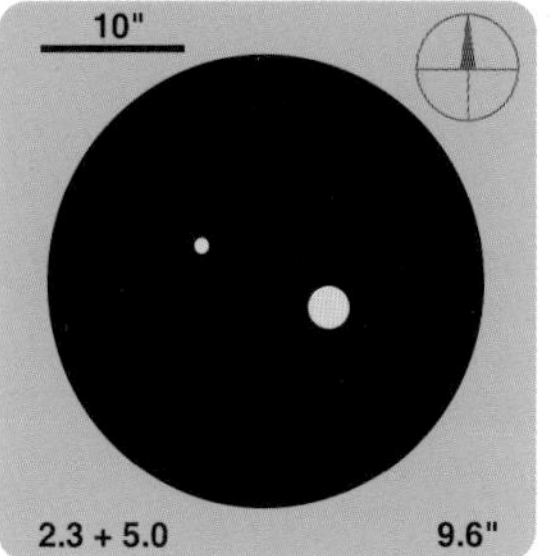

Gamma Andromedae – Alamak, one of the most beautiful colored double stars with orange and blue-green components. Distance: 355 light-years. The fainter component is a very close visual double star (0.6").

ANTLIA

Antlia Ant The Air Pump

One of the fourteen constellations of the southern sky, introduced by the French astronomer Nicolas-Louis de Lacaille after his stay at the Cape of Good Hope in 1750–1754. They are mostly small, indistinct constellations consisting of faint stars. The constellation represents one of the mechanical devices memorialized by Lacaille in the sky. The Air Pump can be found starting from the long shape of the Hydra (the Water Snake).

Alpha Antliae, 4.3 mag, is situated at a distance of some 370 light-years. Its spectral class is K4, which means that it is colder than the Sun. **Upsilon Antliae**, 4.5 mag, is situated at a distance of 700 light-years. It is an orange giant of spectral class K3. **Iota Antliae**, 4.6 mag, is situated at a distance of 200 light-years. Its spectral class is K1.

Zeta 1 and **Zeta 2 Antliae** form an optical double that can be seen even with binoculars. A telescope can discern Zeta 1 as a double, too (see the opposite page).

NGC 3132 (see picture below) is a bright planetary nebula at the boundary of Antlia and Vela. It is situated at a distance of some 2,000 to 2,800 light-years. Its actual diameter is about 0.5 light-year and its angular dimensions are 84" × 52". Its radiation is due to its central star of 10.0 mag and extraordinarily high surface temperature of 100,000 K.

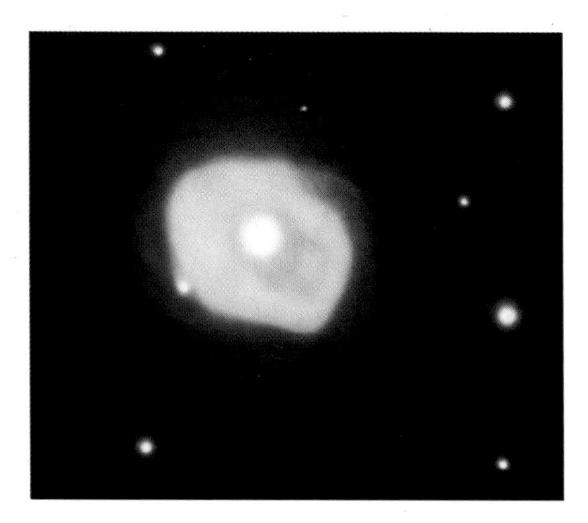

Opposite: **NGC 2997** – a big spiral galaxy of Sc type with two distinct arms. Its brightness is only 11.0 mag. Its beauty can be revealed with a big telescope only.

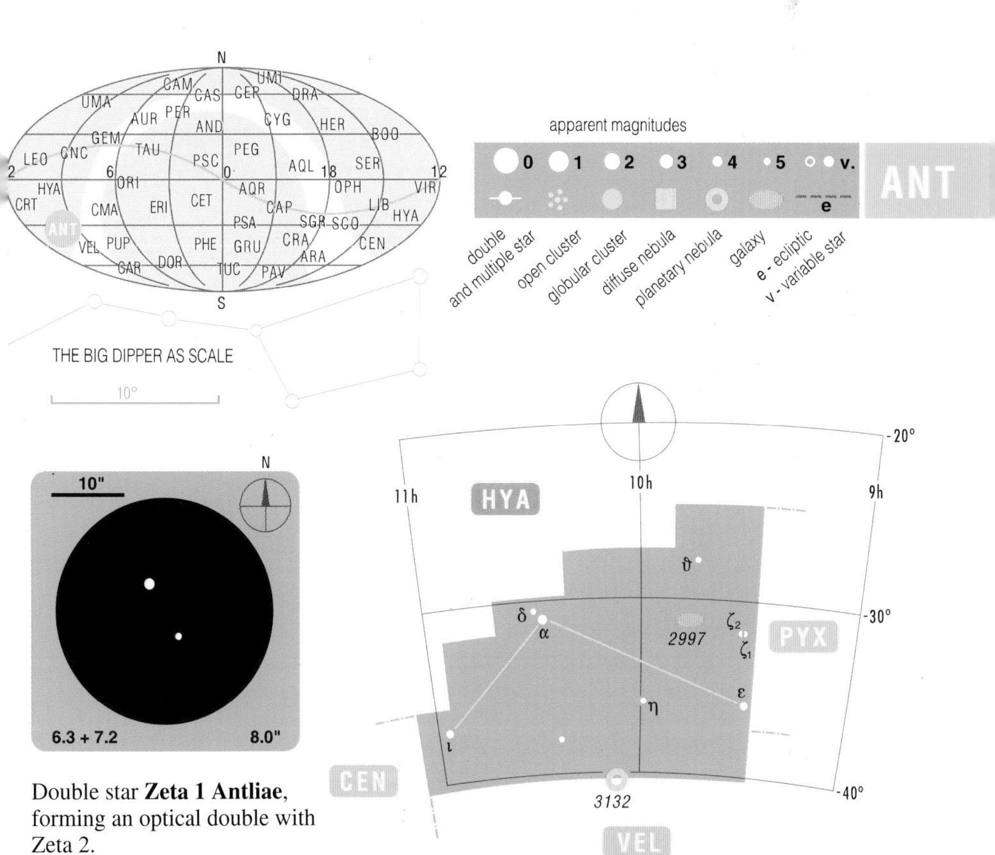

Double star **Zeta 1 Antliae**, forming an optical double with Zeta 2.

APUS

Apodis Aps The Bird of Paradise

One of the twelve southern constellations proposed by the Dutch navigator Keyser at the end of the 16th century and included in Johann Bayer's famous *Uranometria* of 1603. Apus means "footless" in Greek. It was a common practice of the aborigines to cut off the leggs before they sold the birds to Europeans.

When looking for Apus, start from the Milky Way and the Southern Triangle. The picture on the lower right on the opposite page shows the Bird of Paradise as depicted in Hevelius' atlas dating from the end of the 17th century.

Alpha Apodis, 3.8 mag, is the brightest star of the constellation, situated at a distance of 410 light-years. Its spectral class is K5. **Beta Apodis**, 4.2 mag, is a double star with a companion (12.0 mag) at an angular distance of 51"; the brighter component is situated at a distance of 158 light-years and has a spectral class K0. **Gamma Apodis**, 3.9 mag, is a star of spectral class K0, distance about 160 light-years.

Angular Measurements

Astronomical observations cannot get along without knowledge of angular distances and of the angular dimensions of the objects on the celestial sphere. Therefore, let us learn to perform angular measurements in the sky, i.e., the measurements in the degrees of arc! In most cases an estimate made with a sharp eye and an arm as the only instruments is sufficient. If you hold a protractor at arm's length from the eye, every division projected to the sky will correspond to 1 degree of arc (1°). Measure how many degrees are represented by the width of your thumb, fist, spread fingers, and at the nearest opportunity, test your "hand protractor" directly in the sky on the Big Dipper – used as a scale for all constellations in this book. If you're looking at a constellation for the first time, an idea about its angular dimension helps. Small angles can be estimated by comparing them to the diameter of the Moon, which represents 0.5°. You can cover it with your little finger held at an arm's length.

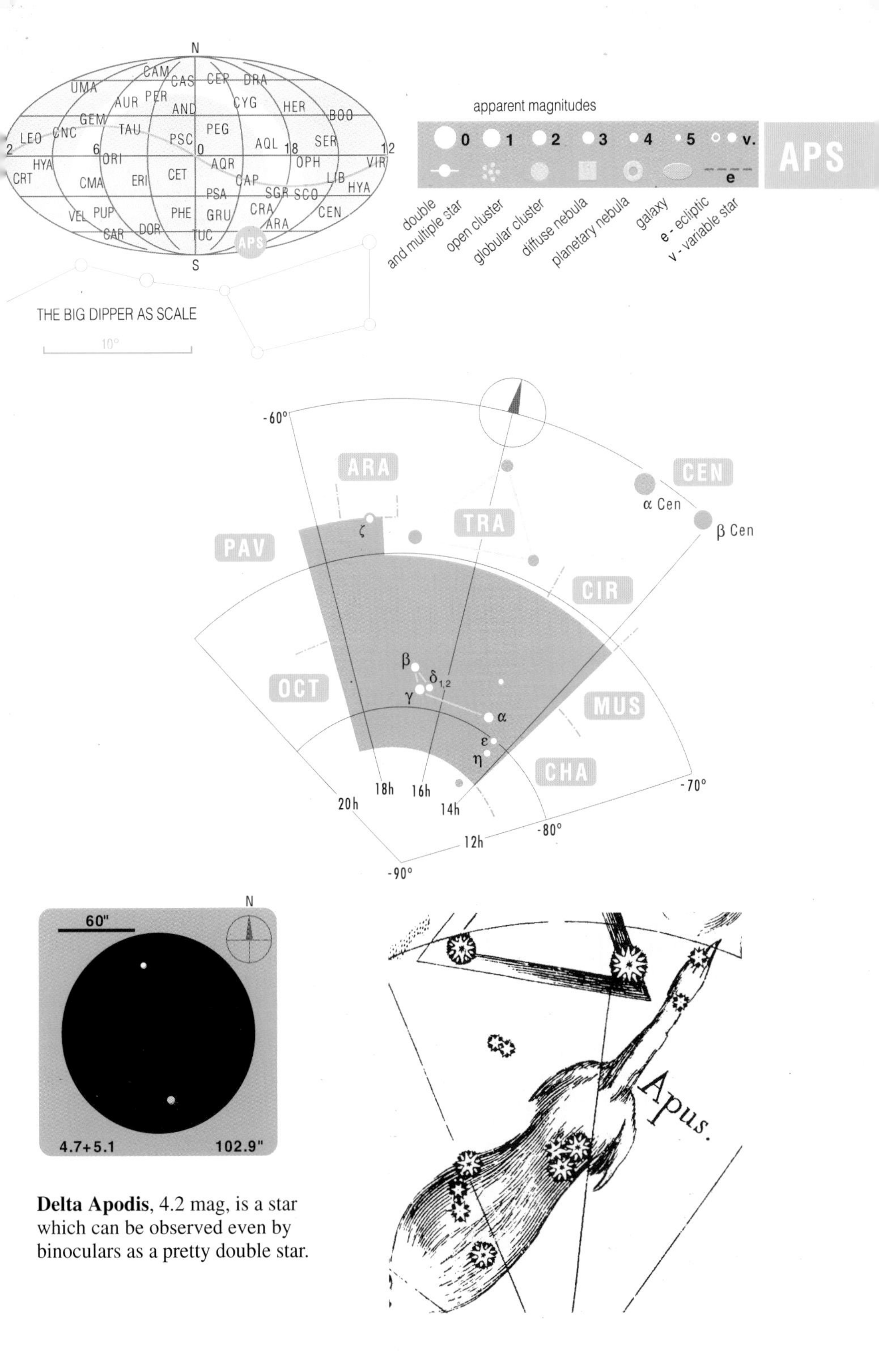

Delta Apodis, 4.2 mag, is a star which can be observed even by binoculars as a pretty double star.

AQUARIUS

Aquarii Aqr The Water Carrier

Zodiacal constellation already known to the ancient Babylonians and Egyptians. The Water Carrier was pictured as a man with a jug – the source of life – giving water. The Water Carrier's jug is a conspicuous group of stars consisting of Eta, Pi, Gamma and Zeta Aquarii, situated on celestial equator south of Pegasus. **Alpha Aquarii – Sadalmelik**, 3.0 mag, is situated at a distance of 760 light-years. Its name derives from Arabic: Al Sad al Melik, means "the lucky one of the king." It is a yellow supergiant of spectral class G2. **Beta Aquarii – Sadalsuud**, 2.9 mag, distance 610 light-years, means "the lucky of the luckiest." It is a yellow supergiant of spectral class G0. **Zeta Aquarii** is a double star with components of 4.3 and 4.5 mag with a separation of 2.0". After 2160 this distance will grow to 6.0". The period of revolution of components is 856 years. The distance of the system is 104 light-years.

M2 – NGC 7089 – bright globular cluster of 6.5 mag and angular diameter of 12'. The distance from the Sun is about 40,000 light-years, actual diameter almost 400 light-years. The cluster contains about 100,000 stars.

M 72 – NGC 6981 – globular cluster of 9.3 mag and angular diameter of 5', actual diameter about 300 light-years. Its distance from the Sun is 55,000 light-years.

NGC 7293 – Helix. The biggest and probably the nearest known planetary nebula of angular dimensions of 12' × 16' (half the Moon's dimensions!) and low surface brightness. In ideal conditions it can be observed even with binoculars. Its distance is estimated between 150 and 450 light-years.

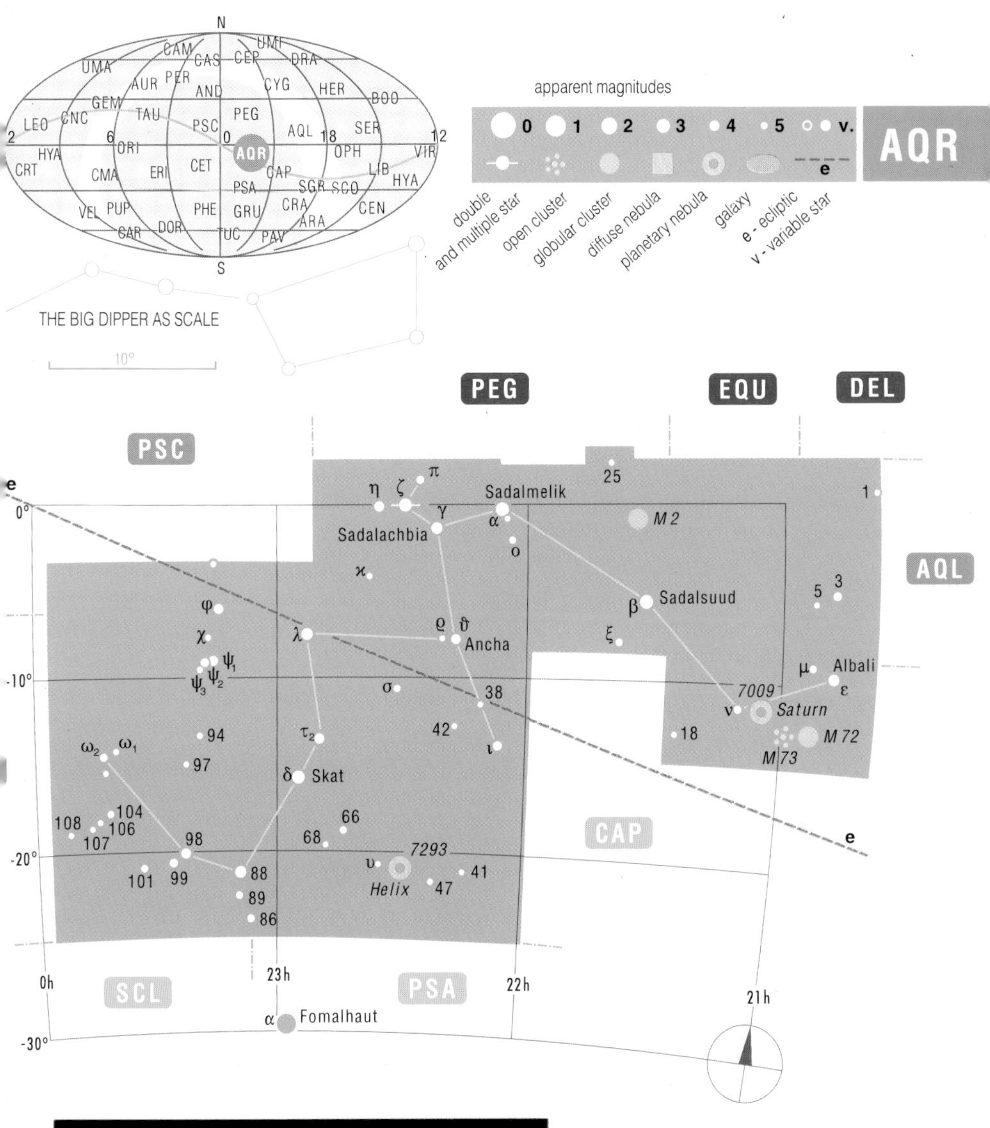

NGC 7009 – Saturn. Bright planetary nebula, 8.0 mag, angular dimension 30" × 26". Its central star is a blue dwarf with a surface temperature of 55,000 K, situated at a distance of some 3,900 light-years.

AQUILA

Aquilae Aql — The Eagle

Constellation known in ancient Mesopotamia and ancient Greece. In mythology the eagle was connected with Zeus, king of gods, whose lightning arrows it used to carry. The bird also brought to heaven the mortal Ganymedes to serve Zeus. The Eagle is a small, beautiful constellation in the Milky Way. Altair, its brightest star, cannot be overlooked. Together with Deneb in the Swan and Vega in the Lyre, Altair forms the "summer triangle."

Alpha Aquilae – Altair is one of the nearest stars, situated at a distance of merely 16.8 light-years. With its magnitude of 0.8 it ranks among the brightest stars. Its spectral class is A7, its diameter equals 1.5 diameters of the Sun, its luminosity is 9 times as high. Altair rotates about its axis at an exceptionally high speed, completing a full revolution every 6.5 hours. Its shape is that of a heavily flattened ellipsoid. **Beta Aquilae – Alshain** is also a near star, situated only 45 light-years from the Sun. It is a yellow giant of spectral class G8. **Gamma Aquilae – Tarazed**, 2.7 mag, is situated at a distance of 460 light-years. It is an orange giant of spectral class K3. It has a companion (10.7 mag) at a distance of 132.6".

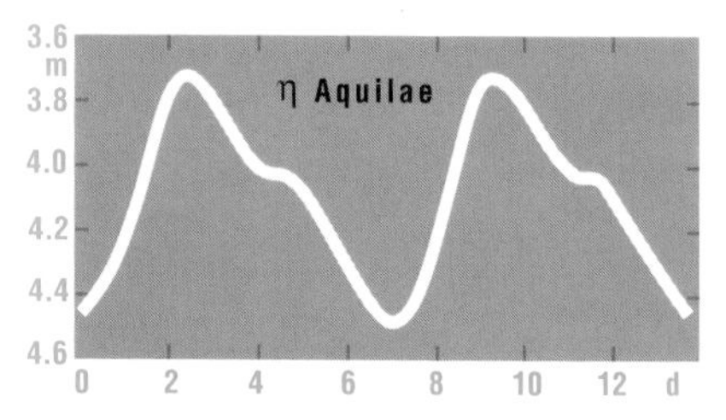

Above: Light curve of variable star **Eta Aquilae**. It is a pulsating variable star – cepheid – situated more than 1,000 light-years away. Its brightness varies by one whole magnitude within a period of 7.2 days.

Below: **NGC 6781** – planetary nebula (11.8 mag), angular dimensions 111" × 109", central star of 16.3 mag.

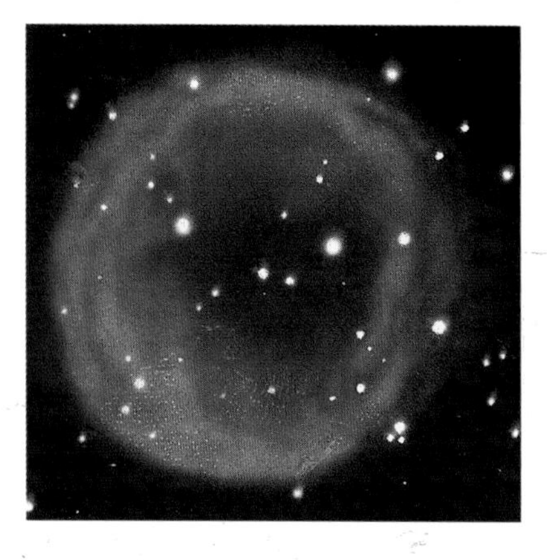

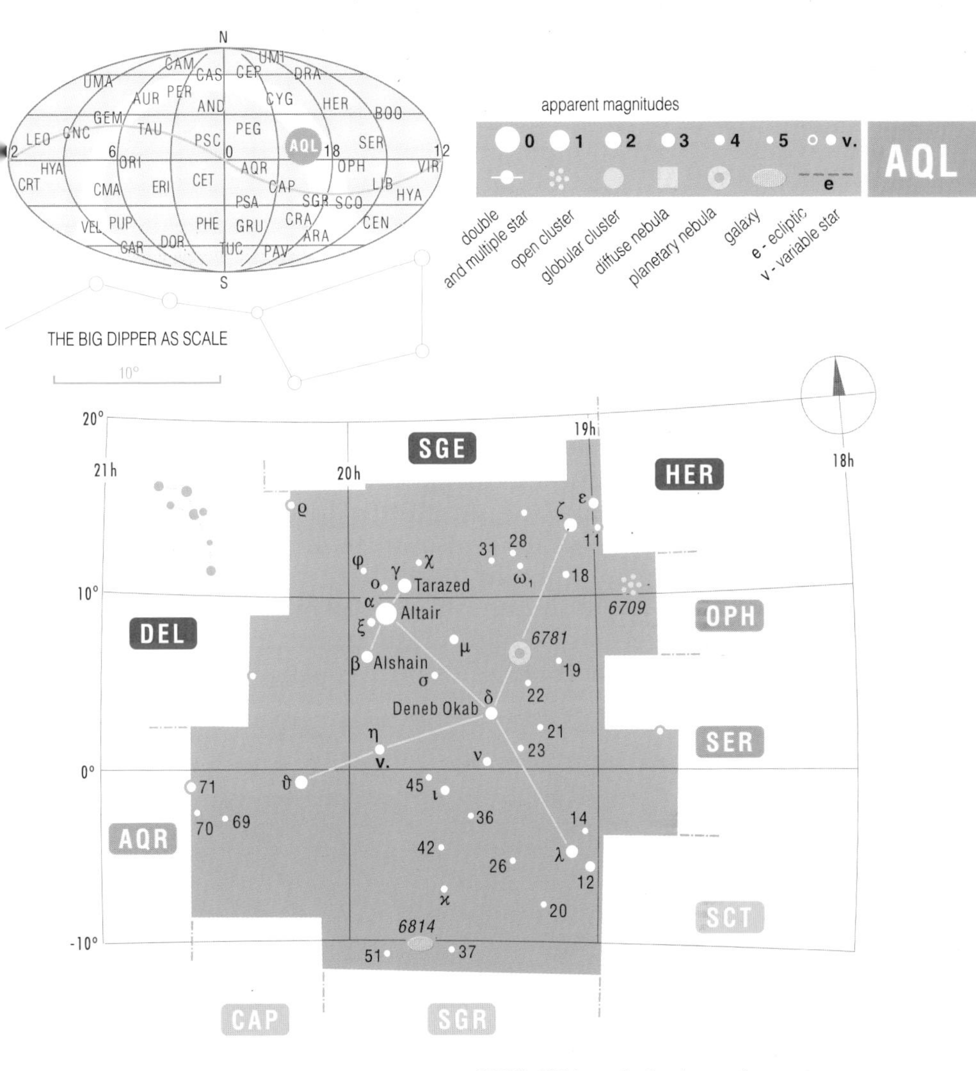

NGC 6709 – bright open star cluster, angular diameter 13'.

NGC 6814 – spiral galaxy of Sb type, angular diameter 3'.

ARA

Arae Ara The Altar

Small but conspicuous constellation recalling a chair or an armchair in shape, known already to ancient Greeks and Romans. In mythology it was the altar of the centaur Chiron, considered the wisest being on the Earth. The constellation was also described as the altar erected by Noah after the flood, as the altar of Moses, the altar from the Temple of Solomon, and others. If you can see from your site at least to the 65th parallel of south declination, the constellation can be found in the Milky Way south of the Scorpion.

The brightest star of the Altar is **Beta Arae** of apparent magnitude 2.9 mag, situated at a distance of 600 light-years. It is an orange giant of spectral class K3. **Alpha Arae**, situated at a distance of 240 light-years, is only one tenth of magnitude fainter. Its spectral class is B2.

The beautiful complex of nebulae connected with the open star cluster **NGC 6193** photographs spectacularly (see below). The emission nebula **NGC 6188** was probably the cradle of this group of young, very hot stars which originated one million years ago. The distance of the object is estimated 4,500 light-years.

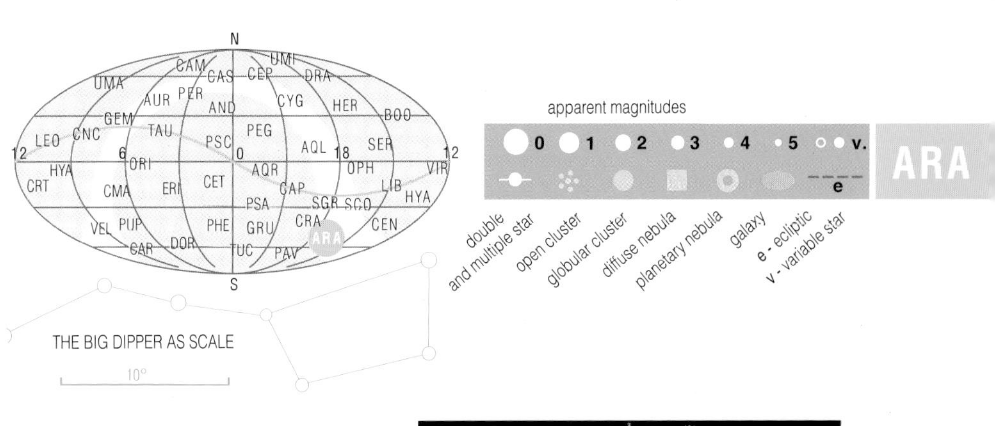

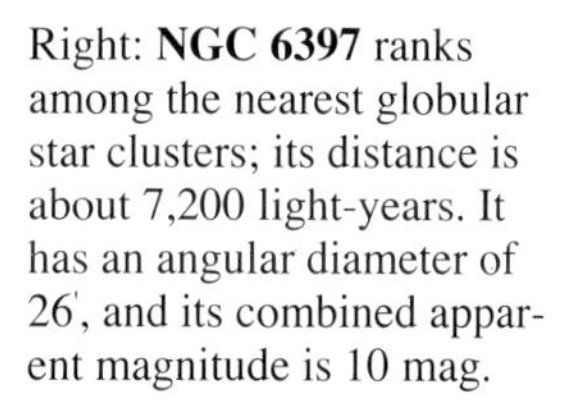

Right: **NGC 6397** ranks among the nearest globular star clusters; its distance is about 7,200 light-years. It has an angular diameter of 26′, and its combined apparent magnitude is 10 mag.

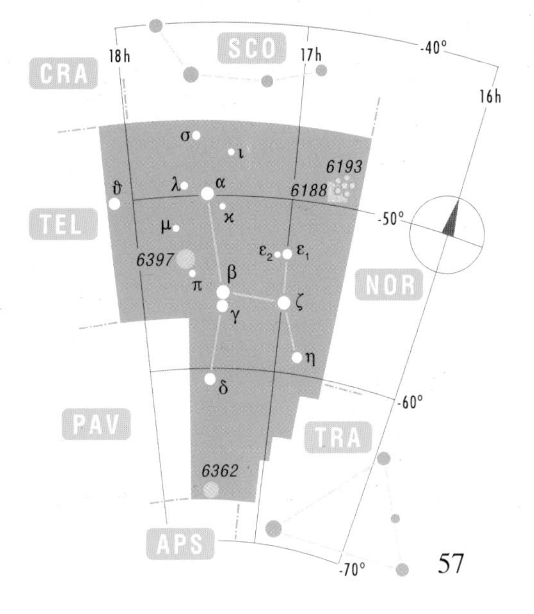

Opposite: In the complex of nebulae around the star cluster **NGC 6193** you can observe the nebula remotely resembling the well known "Horsehead" in Orion. The dark nebula is the remainder of a big cloud of which the cluster originated. Intensive ultraviolet radiation of young giants ionizes and makes visible the gas from the emission nebula **NGC 6188** bordering the dark clouds.

ARIES

Arietis Ari The Ram

Zodiacal constellation, initially situated in the first sign of the zodiac and traditionally related to the arrival of spring. Under the name "The Ram" this constellation was known to ancient Babylonians, Greeks and Egyptians. According to Greek mythology it was the ram with the golden fleece sent by Hermes to save two royal children, Phrixus and Helle. Subsequently the ram was sacrificed to Zeus and its golden fleece was preserved and guarded by a dragon. Finally it was acquired by Jason and his Argonauts who sailed on the ship called Argo.

South of Andromeda our attention is attracted by **Alpha Arietis – Hamal**, 2.0 mag, whose light travels to us for 66 years. It is a giant of spectral class K2. Less than 4° SW from Hamal there is **Beta Arietis – Sheratan**, 2.6 mag, situated at a distance of 60 light-years. The triad of stars characteristic of the Ram is supplemented by **Gamma Arietis – Mesarthim**, one of the best-known double stars, easy to observe even with a small telescope and 204 light-years away.

The symbol of the Ram – ♈ – recalling the ram's horns, is used also as the symbol of the **vernal equinox** situated in this constellation more than 2,000 years ago. At that time the ecliptic was divided into 12 equal sections of 30° each, called the **signs of the zodiac**. The first sector east of the vernal point is still called the sign of the Ram, although it is situated – due to the precession – in the adjacent constellation of the Fishes (Pisces).

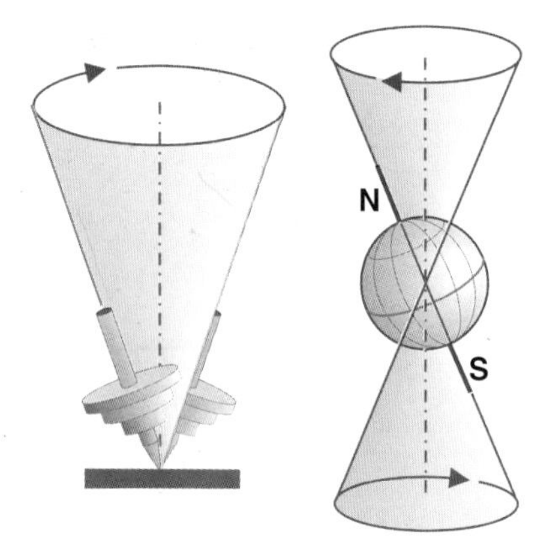

Over the course of 25,725 years (the so-called Platonic Year), Earth's axis of rotation wobbles like a gyroscope. This wobbling motion is called **precession** and is due primarily to the gravitational effects of the Sun and the Moon on the equatorial bulge of Earth. This precession changes the position of the celestial equator among the stars and, consequently, their equatorial coordinates: right ascension and declination.

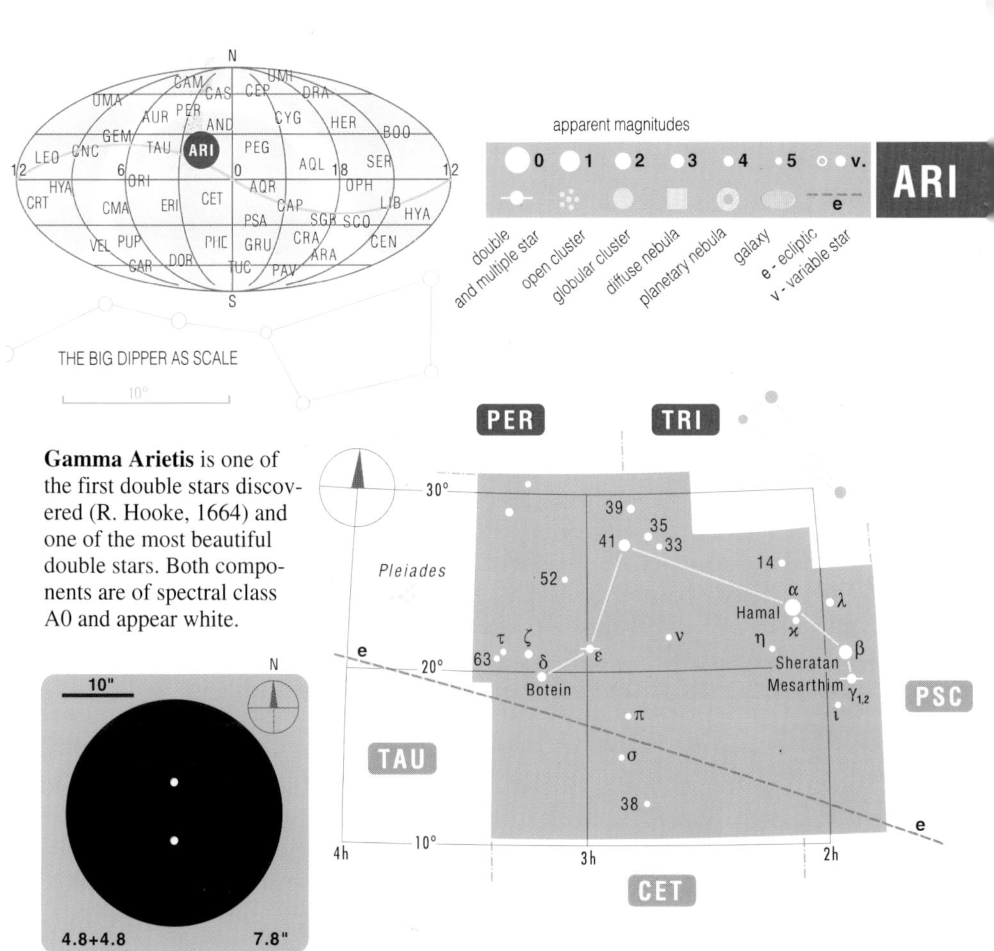

Gamma Arietis is one of the first double stars discovered (R. Hooke, 1664) and one of the most beautiful double stars. Both components are of spectral class A0 and appear white.

Vernal equinox ♈, the point of intersection of the ecliptic **e** with the equator **a**, shifted over the course of the past 2,000 years from the boundary of the Ram far into the Fishes, because of precession. Also the equator **a** changes its position.

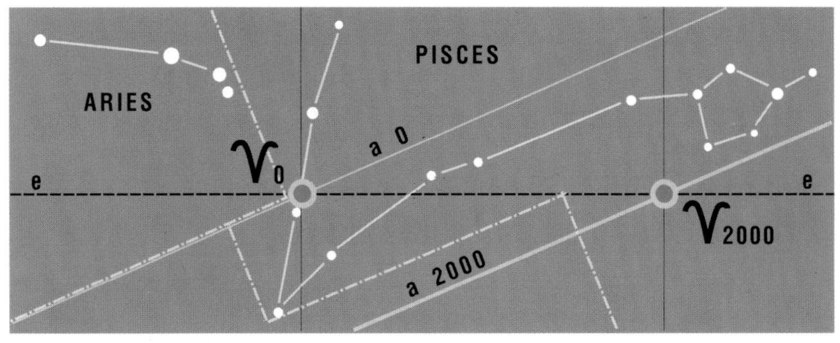

AURIGA

Aurigae Aur The Charioteer

On old maps of the constellations the Charioteer is usually depicted carrying a goat on his back, accompanied by two or three kids. The name of the brightest star of the constellation, Capella, means a small goat, and the kids are the stars Epsilon, Eta, and Zeta Aurigae. According to mythology, the Charioteer was the son of the Roman god Vulcan, Erechtheus, the first man to harness four horses to a chariot.

Alpha Aurigae – Capella is the northernmost star in the "winter hexagon" (0.1 mag, distance 42 light-years). It is a binary consisting of two yellow giants of 9 and 7 Sun diameters respectively that rotate around each other every 104 days. The binary was resolved by a large interferometer at Mount. Wilson Observatory in Pasadena, California, in 1993. **Epsilon Aurigae – Almaaz** is a system of five stars, two of which form a remarkable eclipsing binary, the brightness of which drops from 2.9 mag to 3.8 mag once every 27 years. It is situated near Capella in the sky. During its minimum brightness, the supergiant is eclipsed by its companion, probably surrounded by an enormous cloud of gas and dust.

With the help of binoculars or a small telescope we can find a number of nice open clusters in the constellation. Apart from M 37, mentioned below, other open clusters include **M 36 – NGC 1960** and **M 38 – NGC 1912**, both situated at a distance of over 4,000 light-years. The smaller and brighter M 36 has an angular diameter of 12' and brightness of almost 6.0 mag. The cluster M 38 has a diameter of some 20', apparent magnitude 6.6 mag, and contains about 100 stars. The cluster **NGC 2281**, 6.5 mag, is bright and distinct. Its apparent size equals half of the full Moon's diameter.

Open cluster **M 37 – NGC 2099** is a beautiful object even for a small telescope. Across its apparent diameter of 24', observers have found 150 to 500 and more stars, the brightest of them of about 10.0 mag. The distance of the cluster is about 4,400 light-years, and its actual diameter is about 25 light-years.

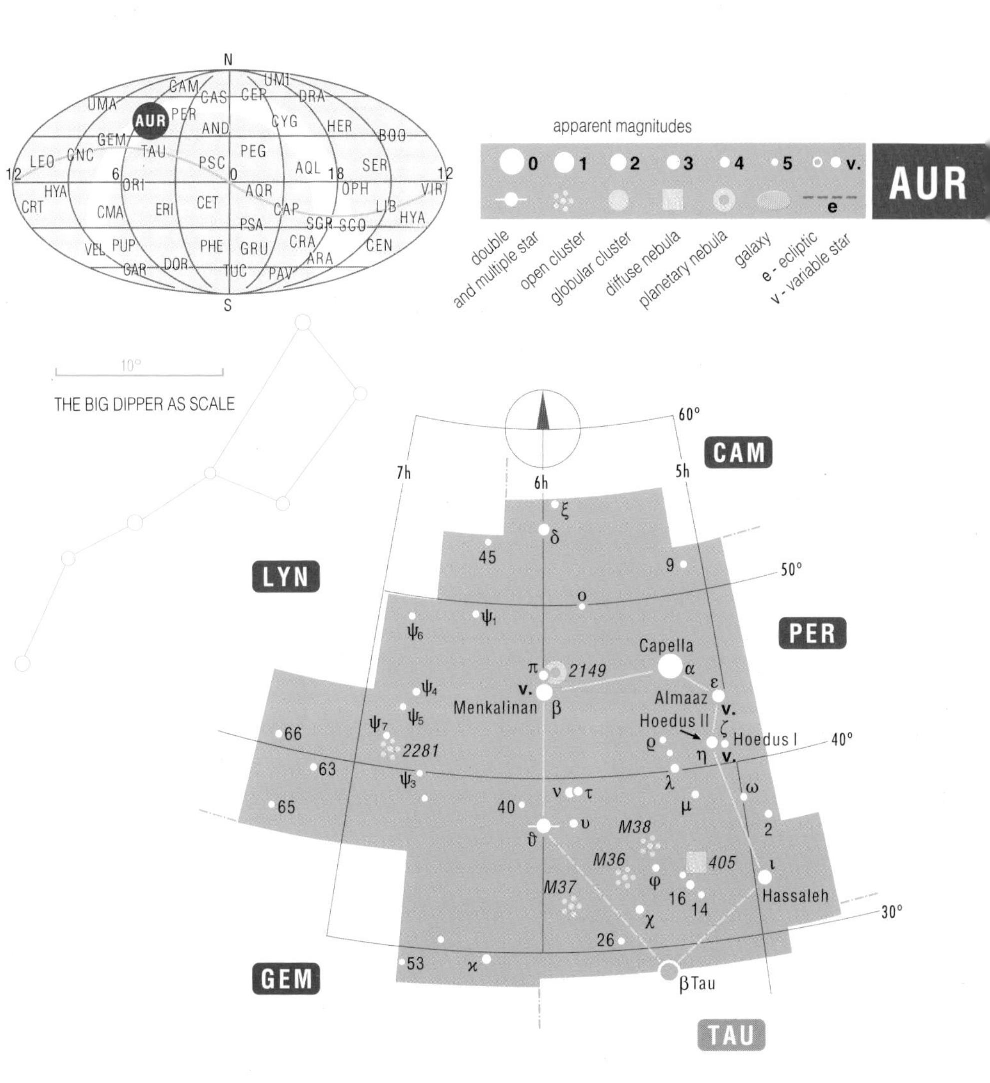

The gas and dust nebula **IC 405** at a distance of some 1,600 light-years is illuminated by ultraviolet light from AE Aurigae (6.7 mag). It is a result of a random meeting of fast-moving star with a cloud of the interstellar matter.

BOÖTES

Boötis Boo The Herdsman

Ancient Sumerians knew the constellation as the Herdsman or the Shepherd. Most outstanding in the constellation is the orange-yellow Arcturus, the fourth brightest star of the sky. The extended arc of the three stars of the tail of the Great Bear (or the handle of the Big Dipper) points toward Arcturus.

The name of **Alpha Boötis – Arcturus** is derived from the Greek arktos – the bear – against which the herdsman guarded his herd. Alpha Boötis is an orange giant of a diameter 27 times the diameter of the Sun – see the picture below. It is helpful to know that Arcturus flies through space at 118 km per second and shifts by 2.28" per year in the sky. Two million years ago, its distance from the Earth was 800 light-years and it appeared near Cepheus as a starlet of 6.7 mag. At present, its distance is 36.7 light-years and its brightness is -0.05 mag. In the next two million years, it will be a faint star of 6.7 mag, situated near the constellation Vela.

In addition to the beautiful multiple stars **Epsilon** and **Mu Boötis** (see below), a small telescope can also show us the double star of **Xi Boötis** with the components of 4.7 and 7.0 mag, separated by 7".

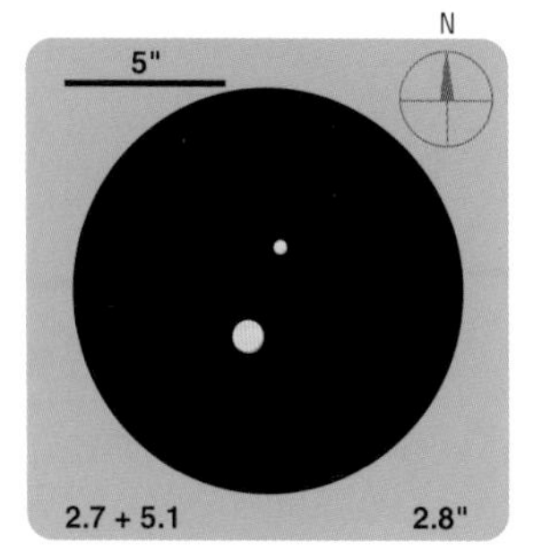

Epsilon Boötis – Izar, one of the most beautiful double stars with a marked color contrast (orange-blue). The pair has been named **Pulcherrima**, from Latin meaning "most beautiful." Its distance is about 210 light-years.

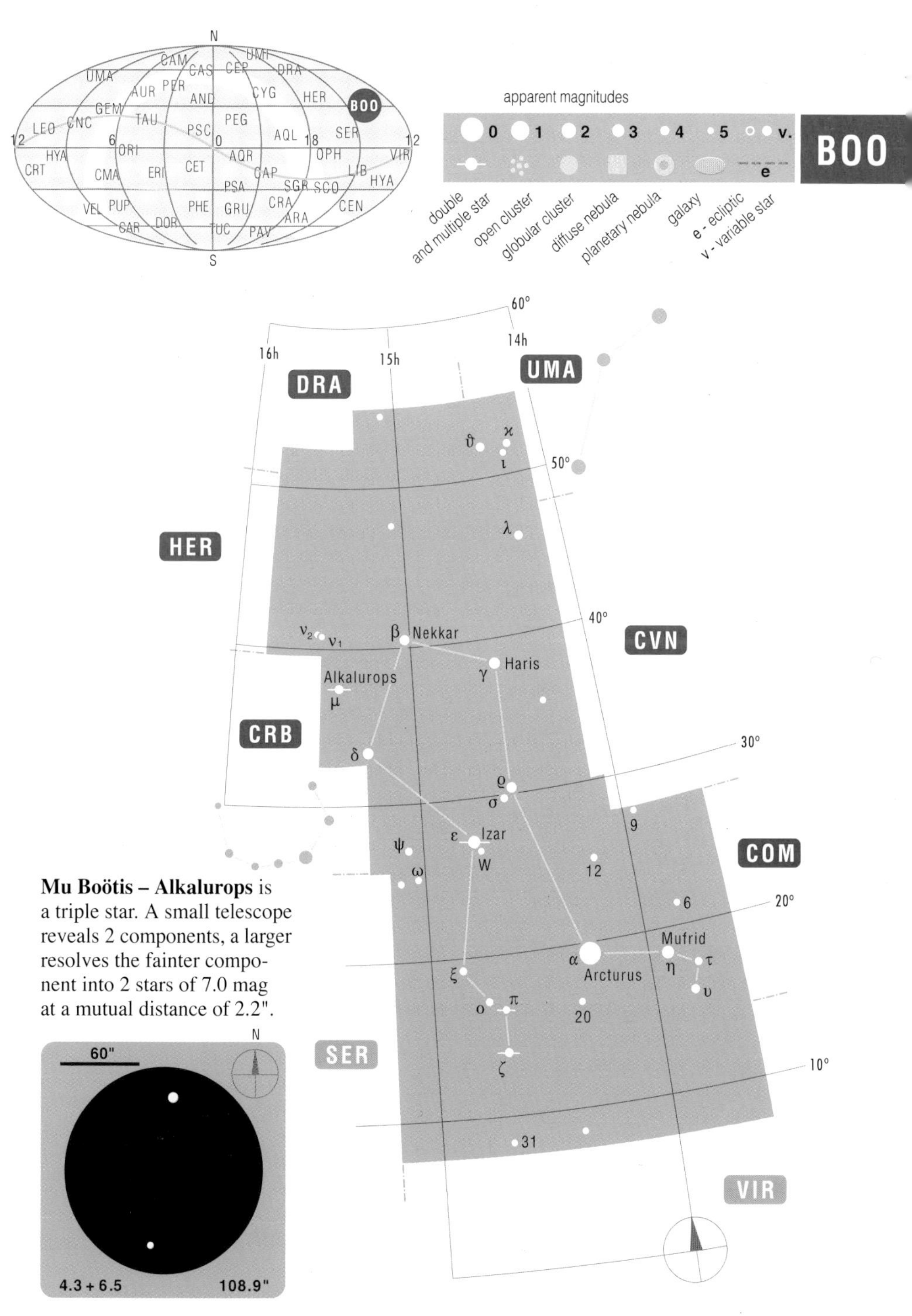

Mu Boötis – Alkalurops is a triple star. A small telescope reveals 2 components, a larger resolves the fainter component into 2 stars of 7.0 mag at a mutual distance of 2.2".

CAELUM

Caeli Cae The Chisel

A small and faint constellation, which is difficult to identify – one of the problematic fourteen constellations introduced by Nicolas-Louis de Lacaille to the southern sky in the 18th century. On older maps it is often denominated as Caela Sculptoris. According to the International Astronomical Union, it is simply called Caelum. Apart from four faint stars of 4.5 to 5 mag, nothing can be observed by the naked eye here. Mark, however, that even an inconspicuous constellation may come in handy some time, for the determination of the position of some celestial phenomenon, for example.

Alpha Caeli, 4.4 mag, is a double star which, however, can be observed by a major telescope only. The companion has an apparent magnitude of only 12.4 mag, its angular distance from the brighter component being 6.6". The light of this system takes 66 years to reach Earth.

Visible and Invisible Radiation

Electromagnetic radiation – and primarily its visible part, light – is the only source of information on deep space objects. The light consists of spectral colors – from violet and blue to yellow and red. We can see radiation of sufficient intensity. Our eyesight, however, can provide only superficial, considerably incomplete information on the space surrounding us. Much more objective and accurate information is provided – in combination with big telescopes – by photographic records or by modern methods of digital picture recording and processing.

Just as important as light for the comprehensive study of a deep space object is its invisible radiation, which can be analyzed by radio astronomy, infrared, ultraviolet, and X-ray astronomy. The records of invisible radiation detectors are often processed into pictures of various colors. There is often an abysmal difference between the printed picture of the object and its actual appearance.

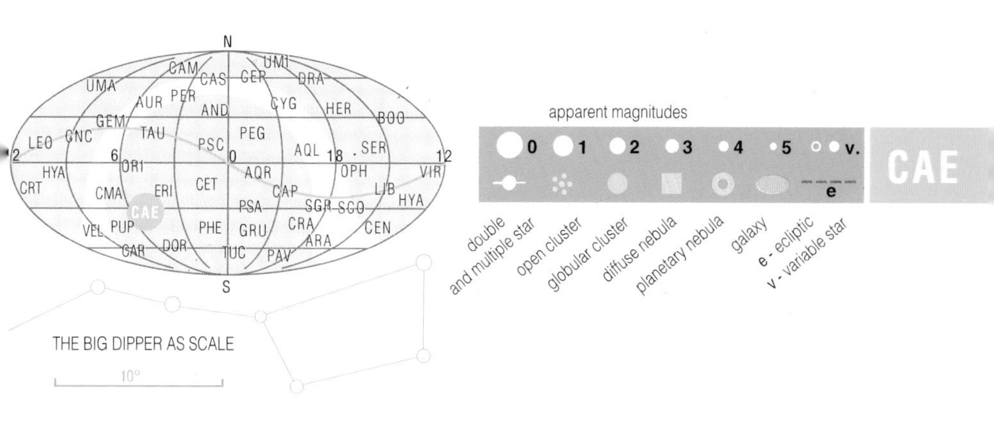

Caelum is situated between the Dove (Columba) and the River Eridanus. It can be found also by means of the nearby Canopus, the second brightest star of the sky. The "meanders" of the Eridanus are a good guide to the orientation of the southern sky.

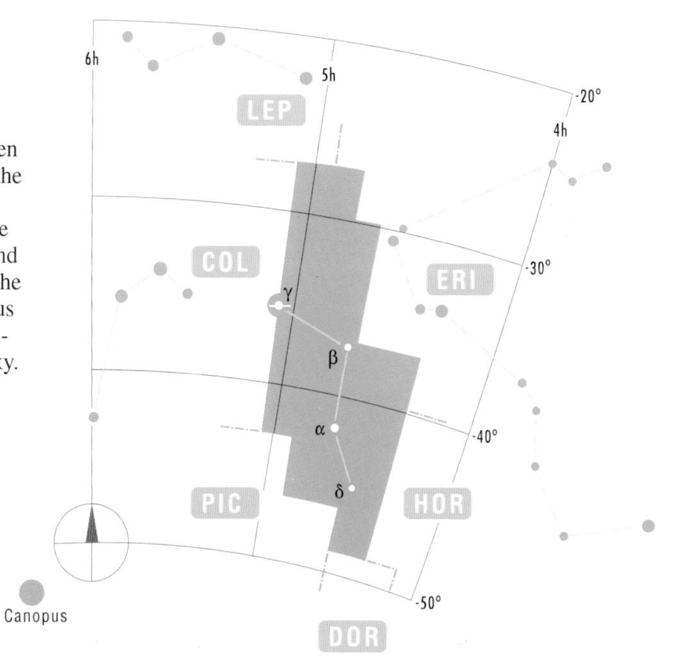

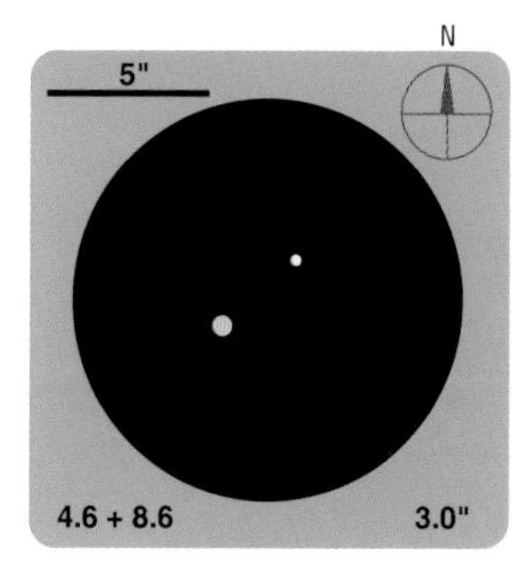

Gamma Caeli can be observed with a small telescope. It is a double star, the brighter component of which is of spectral class K2 and of a very faint yellow-orange hue.

CAMELOPARDALIS

Camelopardalis Cam The Giraffe

The combination of the names for camel and leopard gave rise to the Greek name for the giraffe – an animal with a camel head and leopard spots. The vast area of faint stars between the North Star and the constellation of Auriga was called "The Giraffe" by Jacob Bartsch, the son-in-law of Johann Kepler, in 1624.

Alpha Camelopardalis is a blue supergiant of spectral class O. At a distance of several thousands light-years, it appears as a star of 4.3 mag. **Beta Camelopardalis**, 4.0 mag, a yellow supergiant of spectral class G0, is situated at a distance of 1,000 light-years. It is a double star with a companion of 7.4 mag at an angular distance of 81". Another pretty double star visible in a small telescope is **Sigma 1694 (Struve 1694)** with the components of 5.3 and 5.8 mag separated 21.4".

On the background of the sky without many stars, the open cluster **NGC 1502** of a diameter of 8' stands out, comprising stars of 7.0–10.0 mag.

One galaxy in the Giraffe includes a member of the Local Group, **IC 342**, probably the third nearest galaxy after M 31 and M 33. The abbreviation IC means Index Catalogue, a supplement to the well-known NGC catalogue.

Spiral galaxy **NGC 2403** is the brightest galaxy north of the equator not included in Charles Messier's catalogue. It has an angular diameter of 17.8', total brightness of 8.4 mag, and is of the Sc type. It is some 8 million light-years away, and is probably a member of the Ursa Major group of galaxies together with the big galaxies M 81 and M 82.

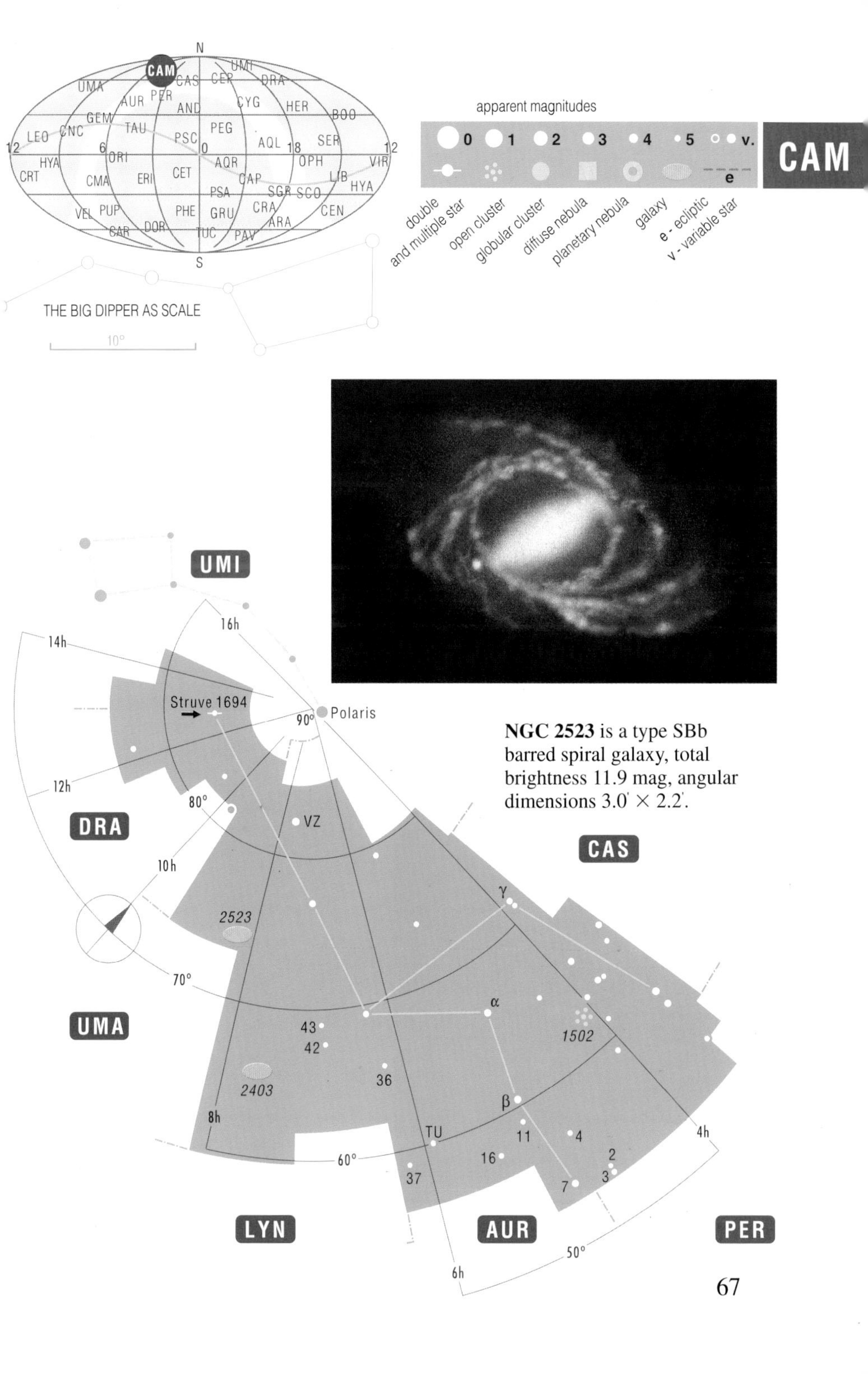

NGC 2523 is a type SBb barred spiral galaxy, total brightness 11.9 mag, angular dimensions 3.0′ × 2.2′.

CANCER

Cancri Cnc The Crab

The least conspicuous constellation of the zodiac. In Greco-Roman mythology, the goddess Hera sent the crab to kill Heracles (Hercules in Roman mythology). The hero stepped on the crab and killed it, but the goddess rewarded the boldness and devotion of the animal by placing it among the constellations. Two or three millennia ago the Sun stood in Cancer on summer solstice, after which it returned south. Cancer is an example of star motion in both directions.

Alpha Cancri – Acubens, 4.3 mag, is a double star which can be observed with a bigger telescope. The companion is separated by 11.3" from the main star and is 1,000 times fainter – its brightness only 11.8 mag. The system is situated some 174 light-years away. The triple star **Zeta Cancri** is shown on the opposite page. The period of revolution of its components A and B is 60 years, their average mutual distance is 19 AU (astronomical units; 1 AU = 150 million km). Component C revolves around them at a distance of 175 AU in about 1,200 years. The distance of this triple star from the Sun is 83 light-years.

M 67 – NGC 2682 is a rich open star cluster of 7.0 mag and angular diameter 28', containing about 500 stars of 9.0 to 16.0 mag. Its distance from the Earth is about 2,300 light-years.

The magnificent open star cluster **M 44 – NGC 2632** has been known as **Praesepe** (the Beehive). The star pair Asellus Borealis and Asellus Australis represents two donkeys (the northern and southern). M 44 can be observed with the naked eye; but binoculars show dozens of more stars of 6.0 mag and fainter. The cluster has an angular diameter of 95' (equal to the diameter of three Moons!). Its distance is about 577 light-years.

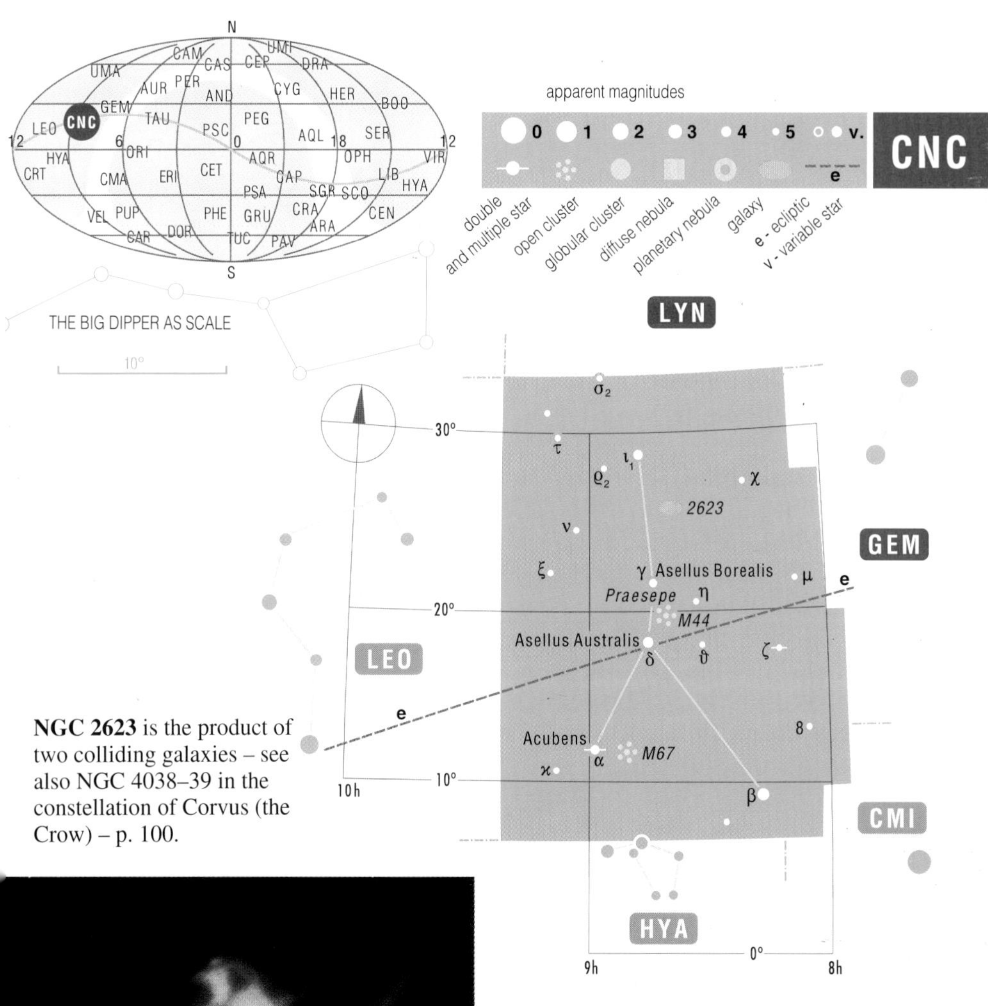

NGC 2623 is the product of two colliding galaxies – see also NGC 4038–39 in the constellation of Corvus (the Crow) – p. 100.

Below: **Zeta Cancri** is a pretty triple star, but its components A and B can be separated only with a larger telescope.

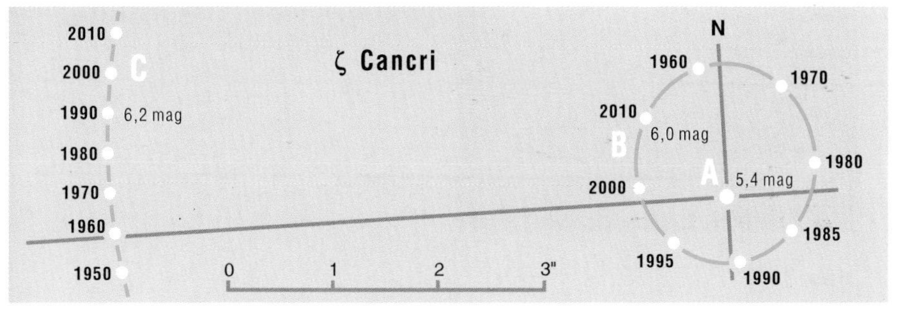

CANES VENATICI

Canum Venaticorum CVn The Hunting Dogs

One of the seven constellations introduced by astronomer Johannes Hevelius in 1690 and included in the list of constellations adopted by the International Astronomical Union in 1930. In celestial maps, it is represented as two hunting dogs with collars and leads held by Bootes (the Herdsman). This inconspicuous constellation south of the Big Dipper contains a number of remarkable objects.

Alpha Canum Venaticorum – Cor Caroli (Charles' Heart) or Chara, 2.8 mag, the star named in honor of Charles I, King of England, in 1660. In a small telescope, it is a very pretty double star (see below). **Y CVn – La Superba** is a bright red variable star, changing brightness from 5.0 to 6.5 mag in a period of 160 days.

M 51 – NGC 5194 – The Whirlpool (or Pinwheel) Galaxy. Its present shape is due to the gravitational effect of the small galaxy NGC 5195 at the end of its longest arm, which approached the big galaxy some 500 million years ago and "pulled" out its long arms. The arms subsequently coiled into the spirals of its present shape (picture on the opposite page).

M 3 – NGC 5272 is a bright globular star cluster (6.0 mag) visible with binoculars. Its diameter in photographs is as wide as 20'. Its distance from the Sun is 34,000 light-years (picture below).

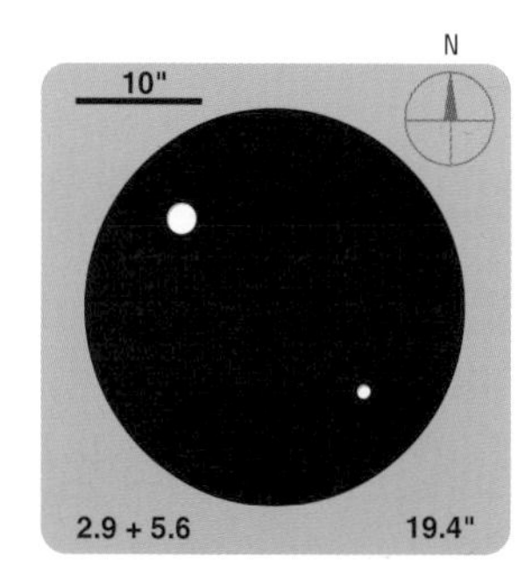

Above: Double star **Alpha CVn**. Its brighter component has a very strong magnetic field and is at a distance of 110 light-years. Its fainter component is 82 light-years away.

Left: Globular star cluster **M 3** contains over 45,000 stars brighter than 22.5 mag. The brightest of them are of 11.0 mag.

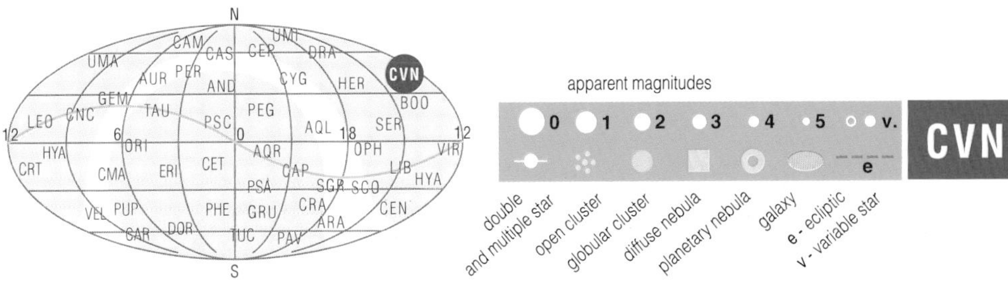

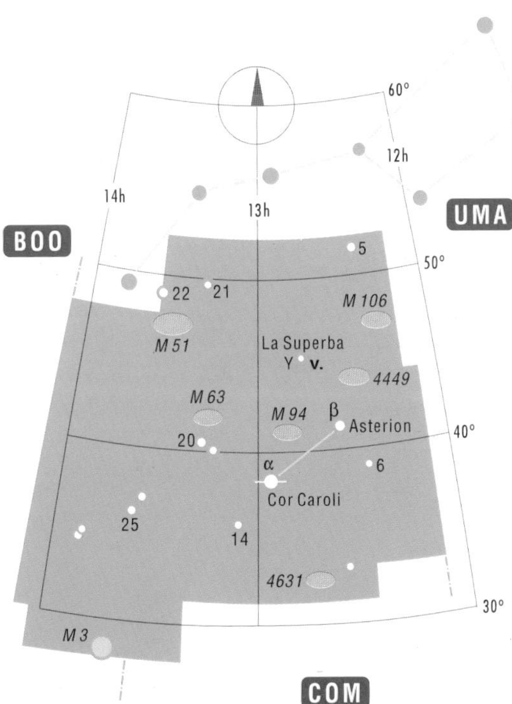

The galaxy **M 51** was discovered by Messier in 1773. Its spiral structure was observed for the first time by Lord Rosse in 1845 with a 72-inch mirror telescope. It is one of the nearest and brightest galaxies of Sc type. Its diameter is more than 100,000 light-years; its distance is about 14 million light-years. NGC 5195, its "satellite" galaxy, apparently connected with the end of the long arm of M 51, is actually situated in the background and partly concealed by the dust clouds in M 51.

CANIS MAJOR

Canis Majoris CMa The Great Dog

Thanks to Sirius, the brightest star in the sky, the Great Dog ranks among the oldest constellations. According to a Greek myth, the Great Dog won a race against a fox, which was the fastest animal in the world. To celebrate his victory, Zeus placed the Dog in the sky. According to another myth, the Great and the Little Dogs accompanied Orion the hunter (p. 160). Orion helps us to find the Great Dog: a line connecting the stars of Orion belt points to Sirius.

Alpha Canis Majoris – Sirius, -1.4 mag, is of spectral class A0. It is 1.8 time as big and 24 times as bright as the Sun. It is actually quite an ordinary star that shines so remarkably only because of its small distance of 8.6 light-years. Sirius' companion (Sirius B) was found by A. Clark in 1862, but it was not until 1915 that it was discovered that Sirius B was a white dwarf whose mass was comparable with that of the Sun, but whose diameter was about the same as that of the Earth and whose density was about 130 kg/ccm. The magnitude of Sirius B is 8.5, but the starlet gets lost next to the bright light of Sirius A. If you want to see it, you need a bigger telescope. After the year 2008, the angular distance of both components will be over 8", which will give amateur skywatchers a chance to see it. Their maximum distance of 11.3" will be attained in 2025.

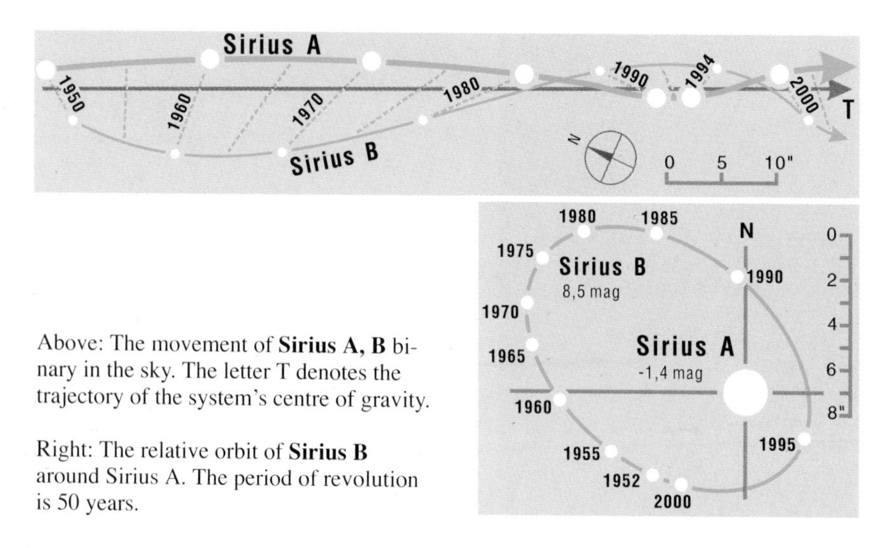

Above: The movement of **Sirius A, B** binary in the sky. The letter T denotes the trajectory of the system's centre of gravity.

Right: The relative orbit of **Sirius B** around Sirius A. The period of revolution is 50 years.

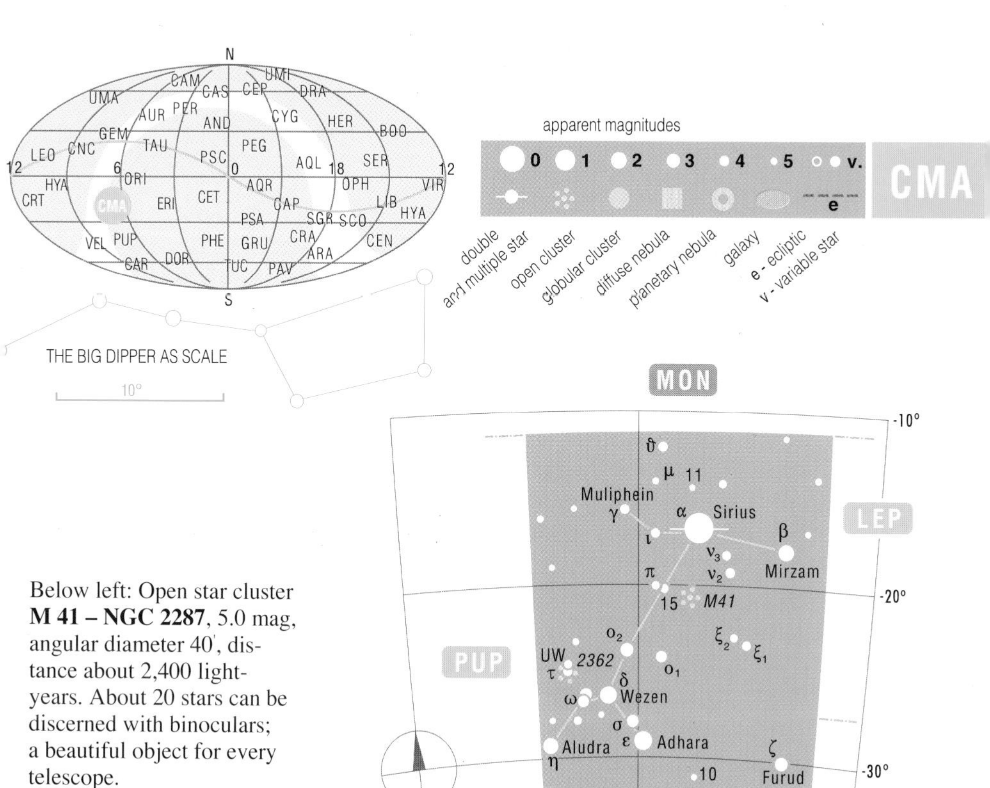

Below left: Open star cluster **M 41 – NGC 2287**, 5.0 mag, angular diameter 40', distance about 2,400 light-years. About 20 stars can be discerned with binoculars; a beautiful object for every telescope.

Below right: Open cluster **NGC 2362**; about 60 stars grouped around the bright star **Tau CMa**, probably not a member of the cluster at all.

CANIS MINOR

Canis Minoris CMi The Little Dog

The smaller of the two dogs accompanying Orion the hunter. Miniature constellation dominated by the bright Procyon, a star of zero magnitude, one of the corners of the conspicuous and almost equilateral "winter triangle" (with Betelgeuse and Sirius in the other corners). The name is of Greek origin, meaning "before the dog." It was so called because the star rises before Sirius and other stars of the Great Dog, if viewed from the Mediterranean and most sites in the northern hemisphere.

Alpha Canis Minoris – Procyon, 0.4 mag, is the eighth brightest star of the sky and one of the nearest. Its light travels to us in 11.4 light-years. Its spectral class is F5, its diameter equals two diameters of the Sun, its surface temperature is 7,000 K, and its luminosity is seven times that of the Sun. Procyon is a remarkable binary star. It is accompanied by a white dwarf of 10.8 mag, fainter than the companion of Sirius. The diameter of Procyon B is about double that of the Earth. White dwarfs are collapsed stars in the final stage of star evolution. They are as big as the planets, but their density is comparable with that of the Sun.

Observing Through the Atmosphere

The atmosphere of the Earth limits and distorts astronomical observation. It acts as a dense filter transmitting the light and radio waves of deep space objects to the surface of the Earth. Other components of electromagnetic radiation are absorbed by the atmosphere at different heights above the Earth's surface and can only be detected by satellites and space probes outside of Earth's atmosphere. However, not even light rays travel from outer space to the surface of the Earth without changes and losses. The radiation is reduced, during its passage through the atmosphere due to diffusion and partial absorption of the light in the atmosphere. This phenomenon is called ***atmospheric extinction*** *and it increases from the zenith to the horizon. The nearer the horizon, the fainter the star appears even if the atmosphere is ideally clean and transparent. The light of the stars is reduced much more by the air pollution, such as dust, smog, or fog. Near cities, observation is limited considerably by the* ***light pollution*** *from street lamps and other artificial light sources.*

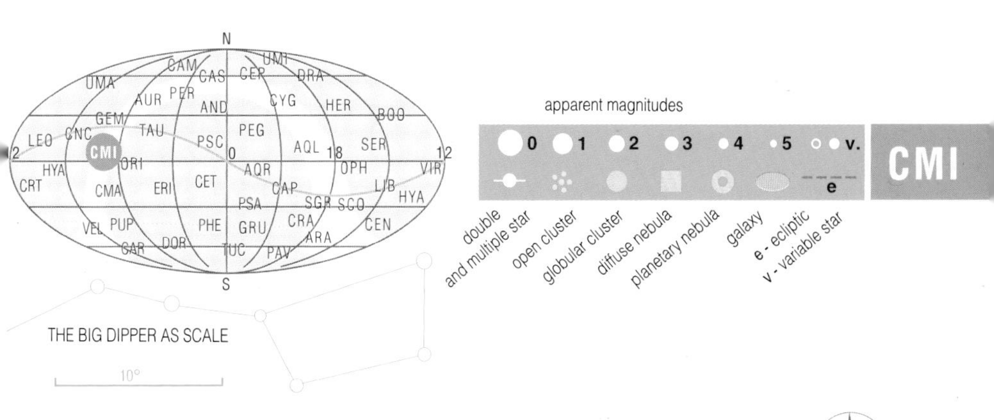

Below left: Relative orbit of Procyon B around Procyon A. The period of revolution is 41 years.

Below right: The 1,000 years' proper motion of Procyon compared with the diameter of the Moon, which is 0.5°. Procyon moves along the celestial sphere at 1.25" per year.

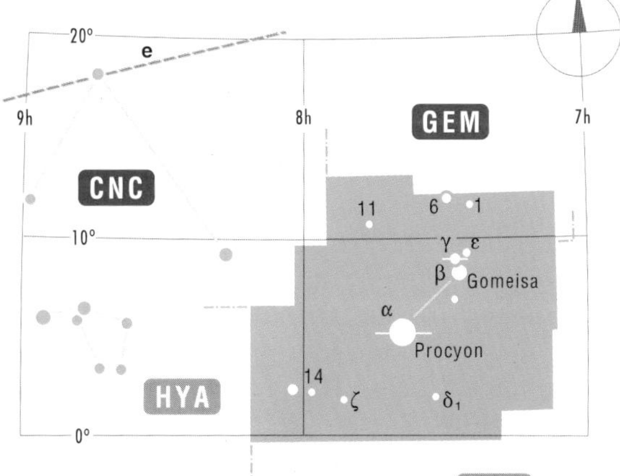

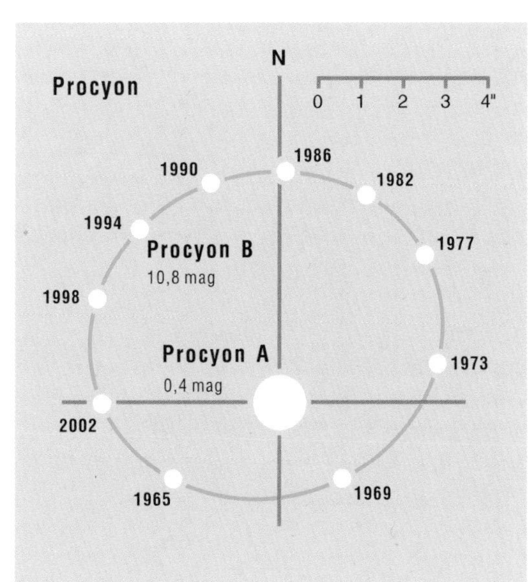

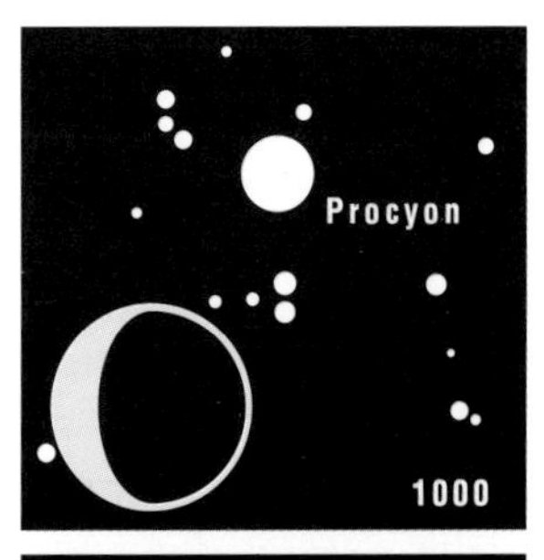

CAPRICORNUS

Capricorni Cap The Goat

Constellation maps show a strange-looking goat with a fish tail (opposite below). An ancient myth tells us that the god Pan was feasting with other gods when suddenly the monster Typhon appeared. To escape it, the gods changed into beasts. Pan, however, in his panicky flight (hence Panic) jumped into the river before changing into a goat entirely, and his lower extremities changed into a fish tail. This metamorphosis was so much to Zeus' liking that he placed the "sea goat" in the sky. This constellation was known to ancient Babylonians. When the ecliptic was divided into signs, Capricorn was in its southernmost part, where the Sun is at the time of the winter solstice – the tropic of Capricorn. Since that time, the constellations of the zodiac are no longer in their original positions due to precession (see also Aries).

Alpha Capricorni – Giedi or Algiedi (see the detail next to the map opposite) is an optical double, the components of which can be discerned with the naked eye in good observing conditions. The stars are actually far apart from one another in space, but happen to lie in nearly the same direction, forming the illusion that they are next to each other – an optical double. The brighter Alpha 2 is situated at a distance of 109 light-years, the fainter Alpha 1 about 690 light-years. Both components are double stars. Globular star cluster **M 30 – NGC 7099**, 7.3 mag, has an angular diameter of 9'. Its distance is about 24,000 light-years.

Atmospheric Refraction

Celestial bodies are not where we see them. Hard to believe? Imagine that the atmosphere of the Earth consists of many thin layers, the density of which decreases with height. A light beam that enters the atmosphere refracts, or changes its direction, slightly at every boundary between the individual layers. Consequently, it reaches your eye from a different direction. The resulting deviation of the beam of light, called ***astronomical refraction****, is lowest around the zenith and highest near the horizon, where it attains 35', more than half a degree. When the Sun or the Moon, the angular diameters of which are about half a degree, touch the horizon with their lower edge, they are actually below the horizon. Temperature differences and movements of atmospheric layers cause quick and irregular changes in the direction of light beams, influencing considerably the quality of observation conditions. It is for the same reasons that the well-known twinkling or shimmering of stars –* ***scintillation*** *– occurs.*

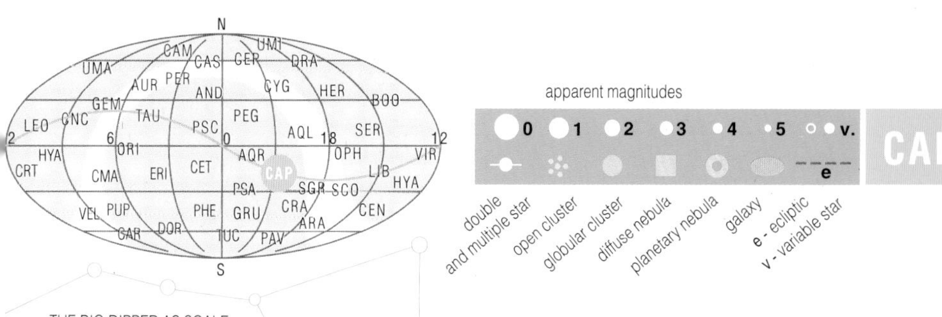

N
S
UMA
CAM
CAS
CEP
UMI
DRA
AUR
PER
AND
CYG
HER
BOO
GEM
TAU
PEG
LEO
CNC
PSC
AQL
SER
6
0
18
12
HYA
ORI
AQR
OPH
VIR
CRT
CMA
ERI
CET
CAP
LIB
HYA
PSA
SGR
SCO
VEL
PUP
PHE
GRU
CRA
CEN
ARA
CAR
DOR
TUC
PAV
THE BIG DIPPER AS SCALE
10°
apparent magnitudes
0
1
2
3
4
5
v.
e
double and multiple star
open cluster
globular cluster
diffuse nebula
planetary nebula
galaxy
e - ecliptic
v - variable star
CAP

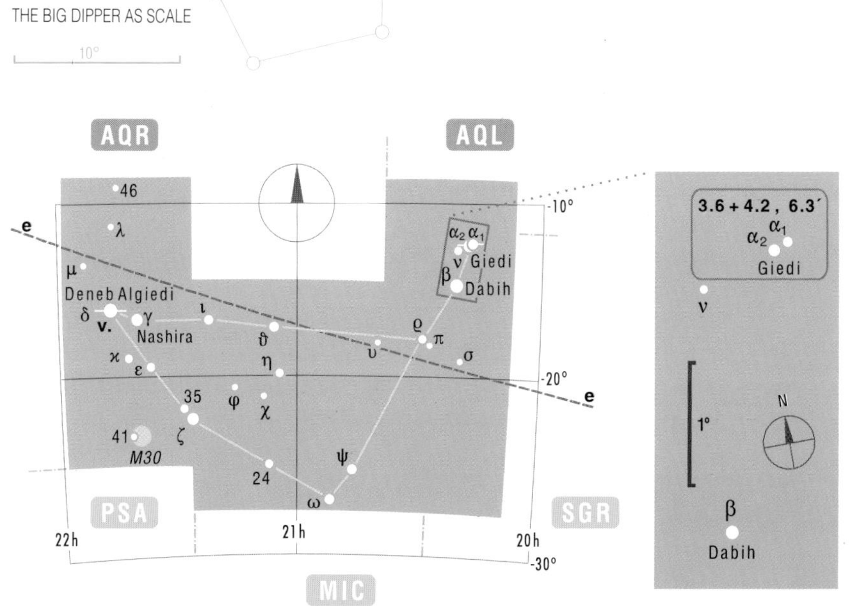

AQR
AQL
46
λ
μ
e
Deneb Algiedi
δ
v.
γ
Nashira
ι
ϑ
η
ϱ
π
υ
σ
ϰ
ε
35
φ
χ
41
M30
ζ
24
ψ
ω
α2 α1
ν Giedi
β
Dabih
-10°
-20°
-30°
PSA
SGR
MIC
22h
21h
20h
3.6 + 4.2 , 6.3´
α1
α2
Giedi
ν
1°
N
β
Dabih

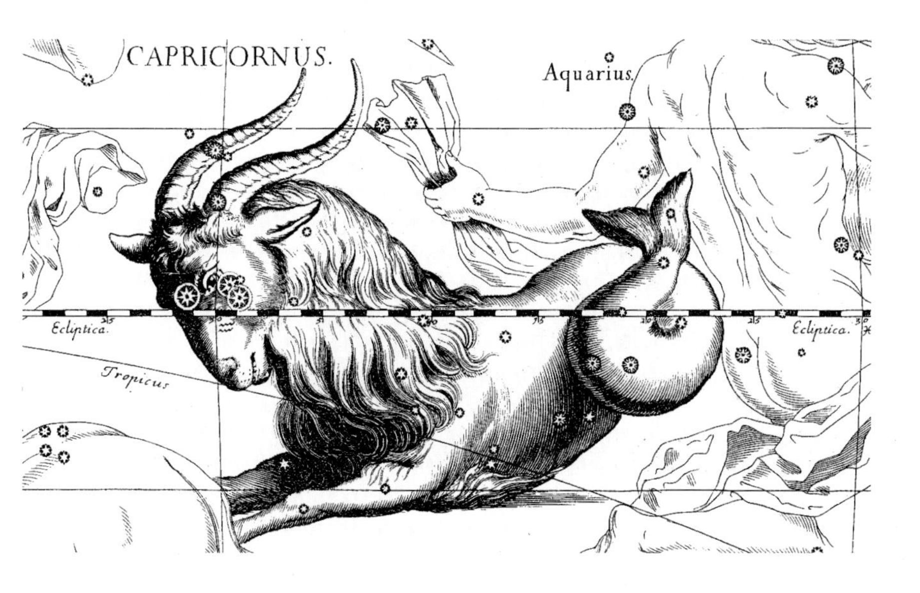

CAPRICORNUS.
Aquarius.
Ecliptica.
Ecliptica.
Tropicus

CARINA

Carinae Car The Keel

Once a part of one of the oldest and biggest constellations of southern sky, "Argo Navis" – the ship Argo that carried Jason and the Argonauts in quest of the Golden Fleece. The constellation comprises two stars of the "false cross," Iota and Epsilon, and the second brightest star of the sky – Canopus. **Alpha Carinae – Canopus**, -0.62 mag, shining brightly from a distance of some 310 light-years, is a supergiant with a diameter of 30 suns. **Upsilon Carinae – Avior**, can be found with a small telescope as a pretty double with components of 3.0 and 6.0 mag, separated 5".

The most remarkable object is **Eta Carinae**, an irregular, nova-like, variable star. During the 17th century its brightness fluctuated between 2.0 and 4.0 mag, increasing to the maximum of -0.8 mag in April 1843, when it became the second brightest star in the sky. In the 20th century, its brightness dropped, varying between 7.0 and 8.0 mag. The star is immersed in the nebula NGC 3372.

NGC 3532 is one of the richest and brightest open star clusters in the sky. The telescope can discern some 150 stars brighter than 12.0 mag. **NGC 2516** is also a bright and distinct open cluster with some 100 stars within an area about one degree across (opposite).

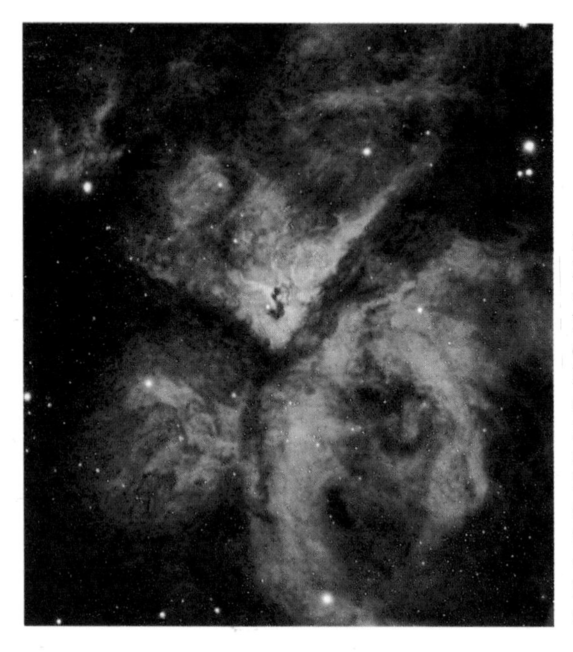

NGC 3372 – The Keyhole Nebula can be observed with the naked eye. It is the brightest H-II region (region of ionized hydrogen) in the Milky Way. The nebula has a diameter of some 400 light-years and is situated at a distance of some 8,000 light-years from Earth. Its apparent diameter is more than 4°. Eta Carinae together with the dark Keyhole Nebula can be found in the middle.

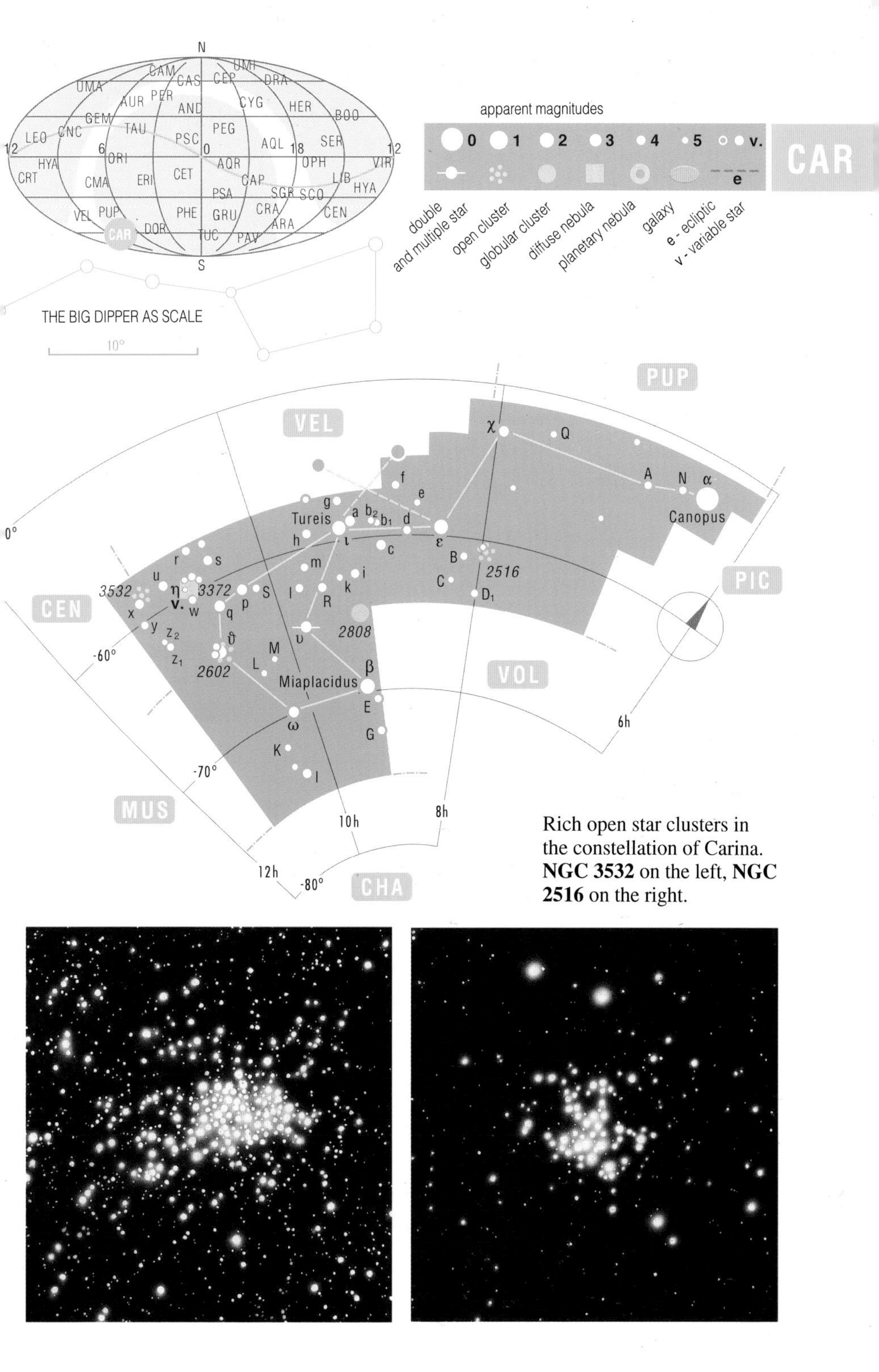

Rich open star clusters in the constellation of Carina. **NGC 3532** on the left, **NGC 2516** on the right.

CASSIOPEIA

Cassiopeiae Cas Cassiopeia

The queen Cassiopeia once claimed that her daughter Andromeda was more beautiful than the sea nymphs, offending Neptune, god of the sea. The heroes of the drama which ensued were immortalized in the form of constellations (see also Cepheus, Andromeda, Perseus, Cetus). Cassiopeia is one of the most prominent and best-known constellations, a group of five bright stars in the form of the letter W. Midway between Cassiopeia and the Big Dipper you can find the North Star – very useful for orientation in the sky.

Gamma Cassiopeiae – Cih, an irregular variable star usually of 2.1 to 2.4 mag. In 1937 it attained 1.6 mag, in 1940 it dropped to 3.0 mag. It is a double star with a companion of 9.0 mag situated at a distance of 2.3" from the main component. In 1572 a supernova in Cassiopeia was observed by Tycho Brahe (then known as **Tycho's Star**), which could be observed with the naked eye for 16 months. The remnant of supernova **SN 1572** is a faint expanding nebula, a source of radio radiation.

Small telescopes can observe several bright open star clusters. **NGC 457** – the Owl Cluster – recalls an owl outlined by stars of 8.0 to 11.0 mag, with "eyes" of 5.0 and 7.0 mag. **M 52 – NGC 7654** contains some 200 stars within a circle of 12' diameter.

The galaxies **NGC 147** and **NGC 185** are distant companions of the Great Andromeda Galaxy M 31.

Open star cluster **M 52 – NGC 7654**.

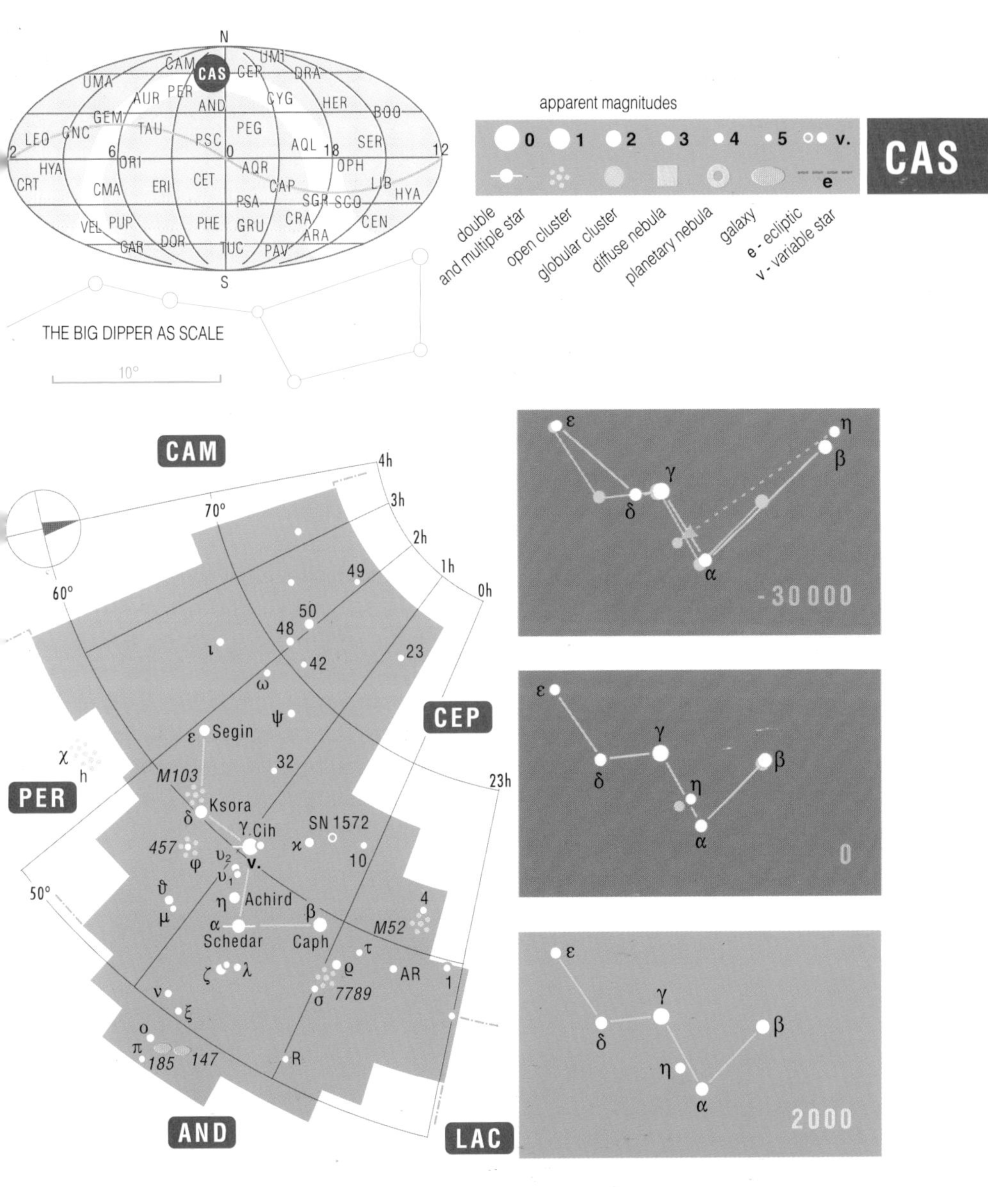

Right: Changes of appearance of Cassiopeia caused by the proper motions of its stars in the period of 60,000 years. The quick-moving **Eta Cas** has passed from one side of the line connecting Alpha and Gamma Cas to the other side since the beginning of our era.

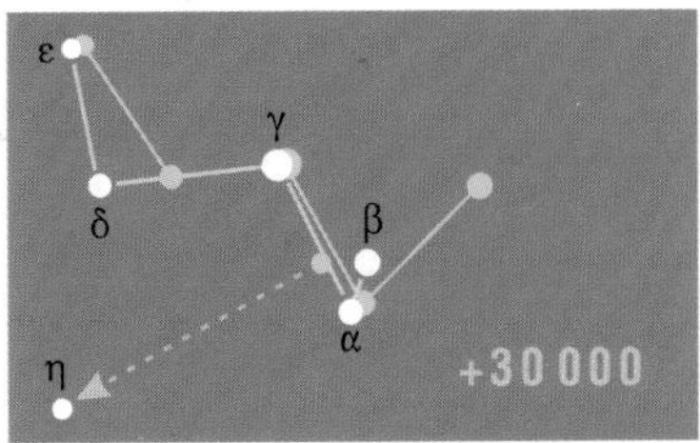

CENTAURUS

Centauri Cen The Centaur

We know centaurs from Greek mythology as strange creatures – half man, half horse. Outstanding among them was the wise centaur Chiron, whom Zeus immortalized as a constellation. The stars Alpha and Beta Centauri serve as "pointers" to the Southern Cross.

Alpha Centauri – Rigil Kentaurus, 0.3 mag, is the third brightest star in the sky, and with its distance of merely 4.4 light-years, the nearest star to the Earth. It is a triple star, its components A, 0.0 mag, and B, 1.3 mag, forming a beautiful double star. The third component C, 11.0 mag, is situated 2°11' southwest of A and B and is nearest to us: 4.249 light-years. Therefore, it is called **Proxima Centauri** (the nearest star of the Centaur). We are not sure, whether it is linked by gravity with the pair of A and B. Proxima was discovered in 1915. It is a red dwarf.

The brightest and the biggest globular star cluster in the sky is **NGC 5139 – Omega Centauri**, visible with the naked eye as a star of 4.0 mag. The distance of the cluster is estimated at 16,500 light-years, its age at about 15 billion years. North of Omega an unusually beautiful galaxy **NGC 5128 – Centaurus A** can be observed with binoculars as a nebulous cloud. It contains three times as many stars as our own Galaxy and its distance is estimated between 10 and 22 millions of light-years. It is an extraordinarily powerful source of radiant energy emitted in optical, radio, infrared and X-ray regions of the spectrum.

Giant globular star cluster **NGC 5139**.

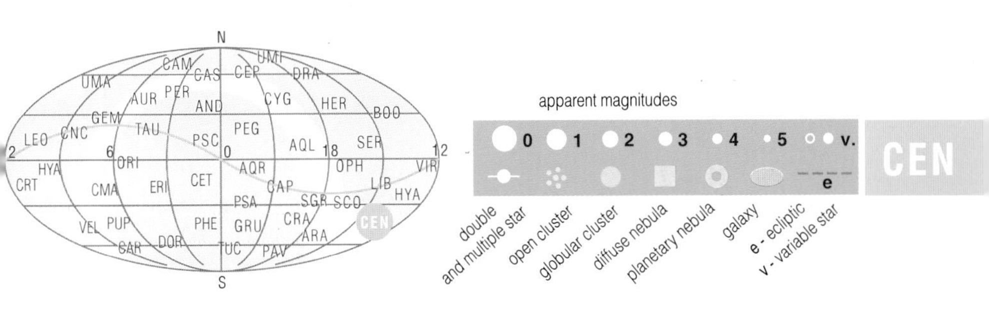

The galaxy **NGC 5128** originated probably by the combination of two galaxies after their collision. The big elliptical galaxy is penetrated by a dark dust ring – the remainder of the other galaxy.

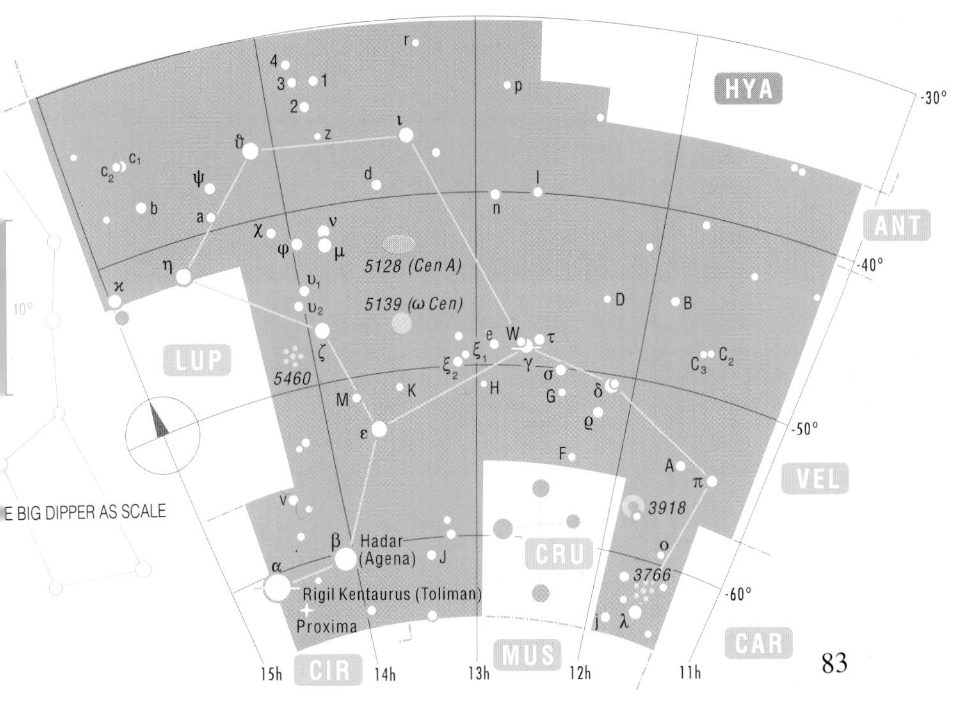

CEPHEUS

Cephei Cep Cepheus

According to Greek mythology Cepheus was the King of Ethiopia, husband of Cassiopeia and father of the beautiful Andromeda. The whole royal family can be found in the sky together with Perseus, the liberator of Andromeda.

The most remarkable object of the constellation is **Delta Cephei**, a pulsating variable star the characteristics and light curve of which are shown on p. 25. It is also one of the most beautiful double stars: its companion (6.3 mag) is situated 41" to the south and looks blue-white next to the golden yellow brighter component. Delta Cephei is the prototype of pulsating variable stars – cepheids (see p. 24); the period of their rhythmical brightening and dimming is proportionate with their luminosity. If we find a cepheid in a distant galaxy, we can measure its period and deduce from it its luminosity and distance.

Another remarkable variable star is **Mu Cephei – the Garnet Star** or **Erakis** of conspicuous red-orange color. Its brightness changes irregularly. The eclipsing binary **VV Cephei** is a red supergiant formerly considered the biggest known star (diameter 1,600 suns). The more recent estimates are much lower.

The galaxy **NGC 6946** is a very faint object (11.0 mag). It belongs to the nearest galaxies beyond the boundaries of the Local Group.

Spiral galaxy **NGC 6946**.

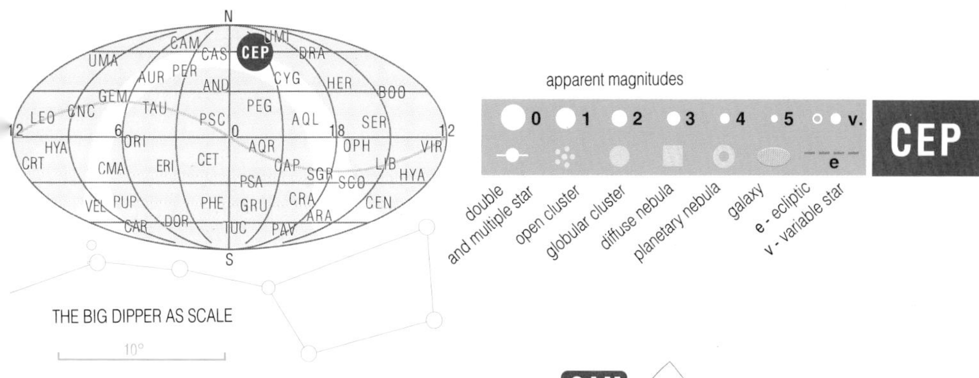

The changing magnitude of **Delta Cephei** can be observed with the naked eye or binoculars, when comparing Delta with two adjacent stars, **Epsilon** (4.2 mag) and **Zeta** (3.4 mag). When in maximum, Delta is approximately as bright as Zeta.

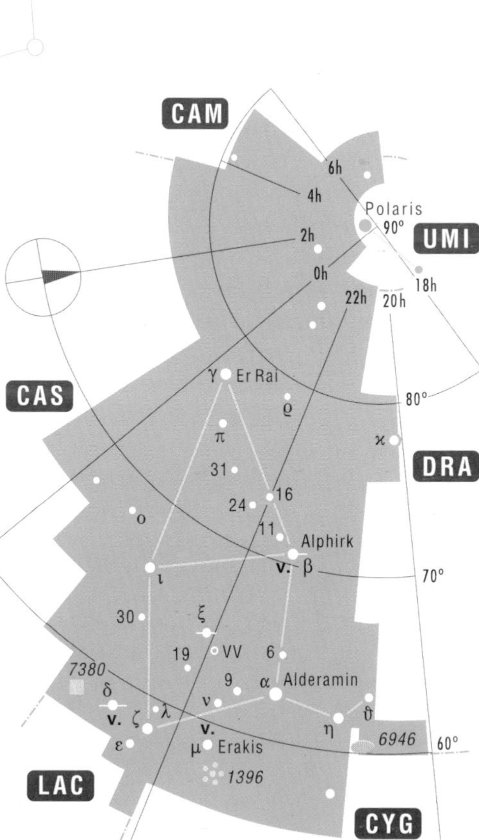

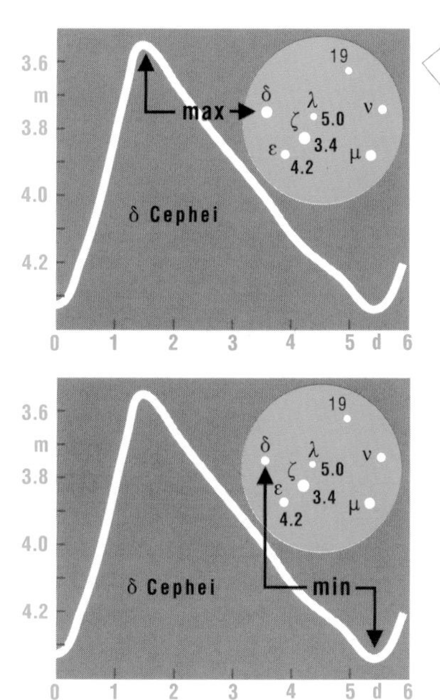

In minimum, Delta Cephei is slightly fainter than Epsilon. Observe them at least 1–2 weeks, regularly every evening. Delta recedes from the maximum to the minimum in 4 days, brightening from the minimum to the maximum in 1.5 days.

CETUS

Ceti Cet The Whale

The fantasy of the creators of pictorial constellation maps has left us pictures of sea monsters with fish tails – which do not at all resemble a whale. Cetus was the sea monster intended to devour Andromeda, and was killed by Perseus at the last moment. In Latin, however, cetus means the whale. According to a more recent legend, it is supposed to be the "big fish" that swallowed Jonah (The Old Testament). Apparently, the world of myths and legends does not mind that a whale is not a fish, but a mammal.

Omicron Ceti – Mira (from the Latin for "wonderful") is the brightest and best-known long-period variable star. Its brightness changes approximately from 3.0 mag to 9.0 mag in 332 days on average. Mira is a red supergiant with a diameter of some 300 suns. Its distance from the Earth is 420 light-years. **Gamma Ceti – Kaffaljidmah** is a double star with the components of 3.5 mag and 6.2 mag separated at 2.8". **Tau Ceti**, 3.5 mag, situated at a distance of merely 11.9 light-years, ranks among the 17th or 18th nearest star to our solar system. Since 1959 the signals of its hypothetic civilization have been sought, but none have been ascertained.

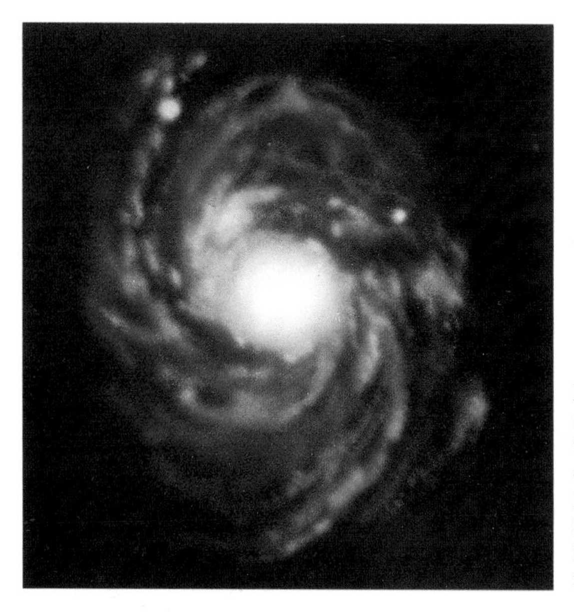

M 77 – NGC 1068 is a spiral galaxy of SB type, 9.0 mag, at a distance of over 50 million light-years. It belongs to the so-called Seyfert galaxies with very intensive radio radiation from the nucleus.

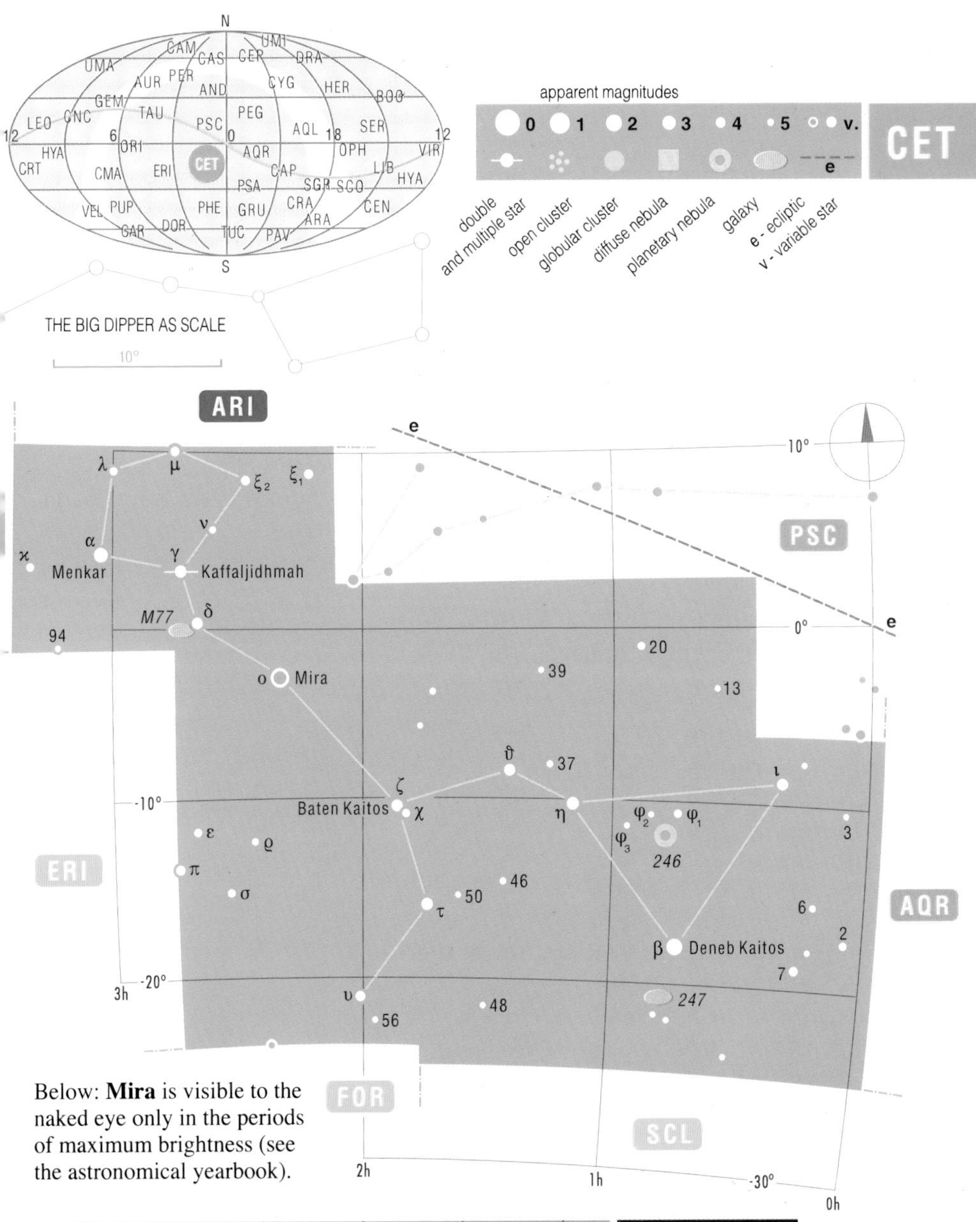

Below: **Mira** is visible to the naked eye only in the periods of maximum brightness (see the astronomical yearbook).

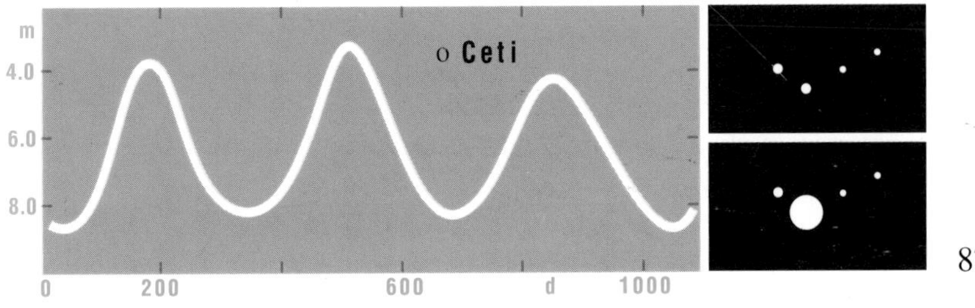

CHAMAELEON

Chamaeleontis Cha The Chameleon

In nature the chameleon is known for its ability to make itself inconspicuous. The navigator P. Keyser succeeded in molding several faint stars into a very inconspicuous chameleon. J. Bayer presented the new constellation in his *Uranometria* in 1603. The small and hardly perceptible constellation is almost lost in the region devoid of brighter stars near the south celestial pole. It can be found by means of the Southern Cross, the longer arm of which points to the south pole across the Chameleon.

The brightest star of the constellation is **Alpha Chamaeleontis**, 4.1 mag, a white giant of spectral class F5 from which light travels to us 64 years. **Beta Chamaeleontis**, 4.2 mag, is of spectral class B5 and is situated on the main sequence of the temperature-luminosity diagram. Its distance is 270 light-years. **Gamma Chamaeleontis**, 4.1 mag, an orange-red giant of spectral class M0, shines from the distance of some 410 light-years.

Planetary nebula **NGC 3195** is a faint object which appears as a nebulous sphere with a dimension of 30" × 40" – about as large as Jupiter.

How Many Stars Can We See in the Sky?

Poets speak about billions, because exaggeration is their prerogative. The reality is far more modest. On a moonless night under good observation conditions (transparent atmosphere, no artificial lighting, etc.) we can count some 2,000–3,000 stars above the horizon. Their number depends on the geographic latitude of the observation site, the distribution of the stars on the celestial sphere and atmospheric extinction (p. 74), so that the numbers of stars visible to the naked eye may vary substantially from the average. If you have perfect eyesight, you can see some 6,000 stars in the whole sky.

The number of visible stars increases many times if we use a telescope. A very small telescope or binoculars can show some 100,000 to 200,000 stars, big instruments show millions to hundreds of millions of stars, as well as deep space objects, like galaxies (see also p. 179).

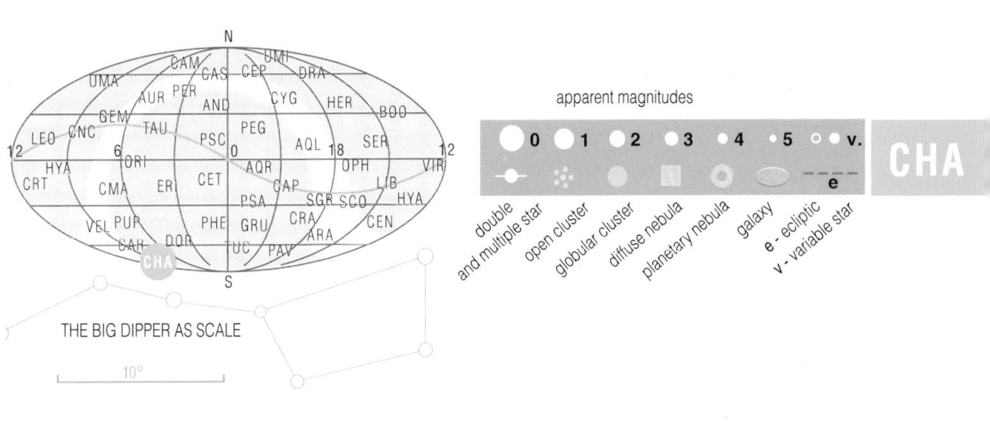

The optical double star **Delta 1 Cha**, 5.5 mag, and **Delta 2 Cha**, 4.5 mag, can be separated with binoculars. The distance of Delta 1 is 354 light-years, that of Delta 2 is 364 light-years. **Delta 1** (below) is a binary the components of which can be discerned with a big telescope only.

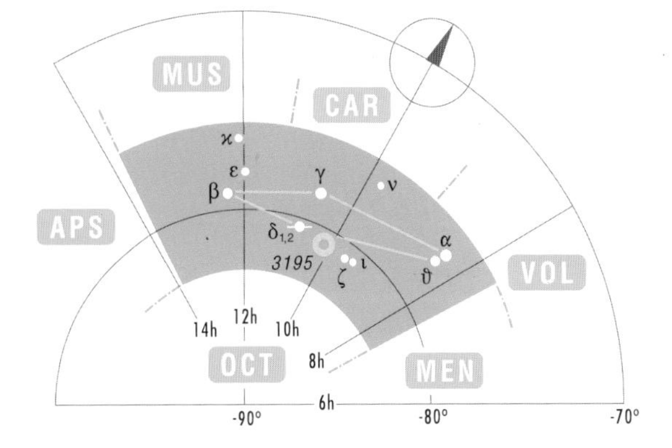

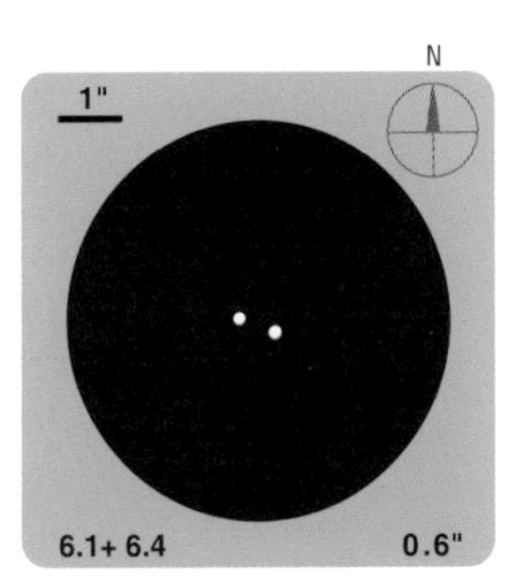

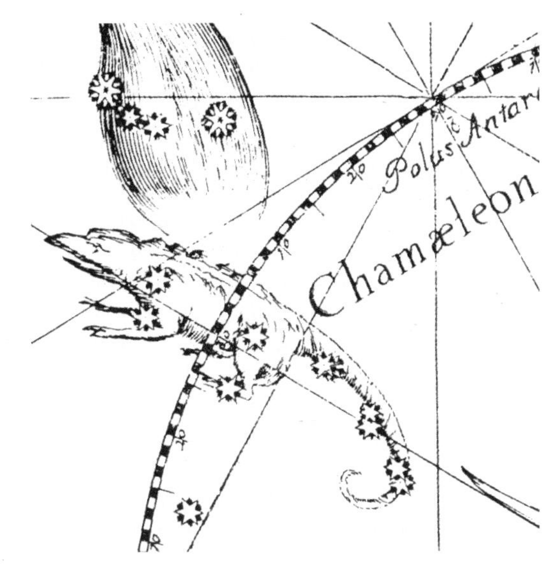

The picture of Chameleon in the stellar atlas of J. Hevelius' *Uranographia* from 1690.

CIRCINUS

Circini Cir The Compasses

One of the smallest "modern" constellations with which the French astronomer Nicolas-Louis de Lacaille (also known as Abbé de la Caille, 1713–1762) filled every free area in the southern sky in the middle of the 18th century. Lacaille headed the expedition of the French Academy of Sciences to the Cape of Good Hope in 1750–1754 to observe the southern sky and to map it systematically for the first time. The detail of his map showing the constellation the Compasses (Le Compas) is on the opposite page. The Compasses can be found relatively easily as it is situated in the immediate proximity of the conspicuously bright stars Alpha and Beta Centauri.

Alpha Circini, 3.2 mag, and **Gamma Circini**, 4.6 mag, are double stars (shown on opposite page). The distance of Alpha is 54 light-years, that of Gamma 500 light-years. **Beta Circini**, 4.1 mag, is a star of the main sequence, of spectral class A3, at a distance of 97 light-years.

Star Colors

A cursory glance at the sky tells us that all stars are white. A more careful observation, however, will reveal a whole number of very subtle color hues, particularly in the case of brightest stars. Here are a few examples:

- *light orange: Betelgeuse, Antares, Aldebaran*
- *light orange-yellow: Arcturus, Pollux*
- *yellow: Capella, Alpha Centauri*
- *blue: Spica, Regulus, Vega*
- *white: Deneb, Altair, Procyon*

Astronomical literature generally gives color as a characteristic of the thermal radiation of the star and its spectral class. For instance, cold stars (below 5,000 K) are described as orange (class K), red, and crimson (class M).

We perceive a distinct difference of color in the case of double stars where, for instance, the fainter component (thanks to an illusion due to contrast) may appear as a complementary color to its brighter component of the same spectral class.

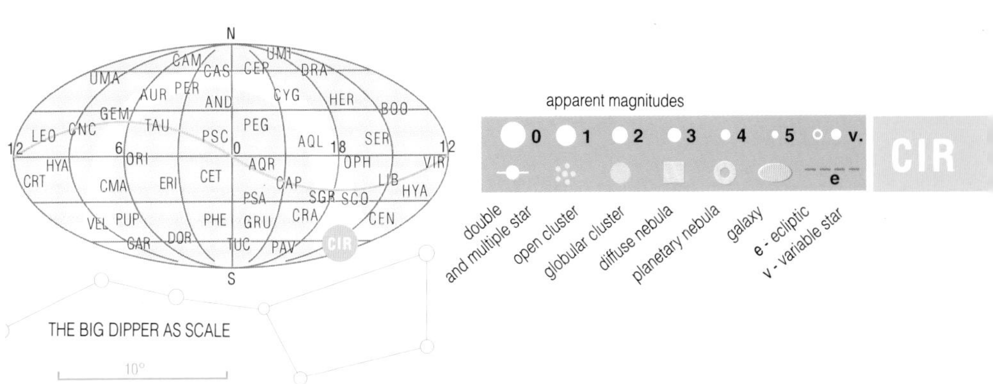

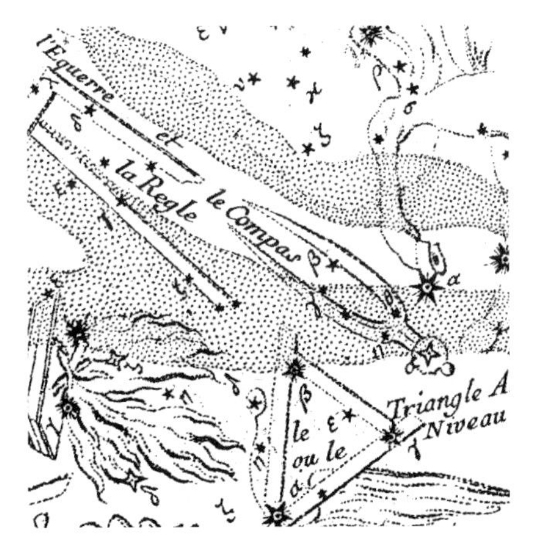

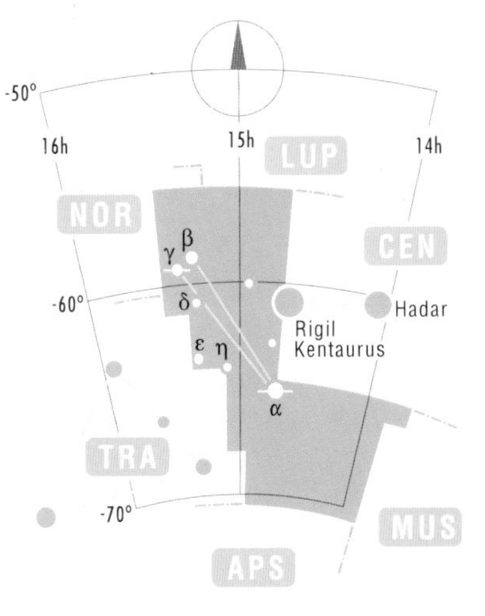

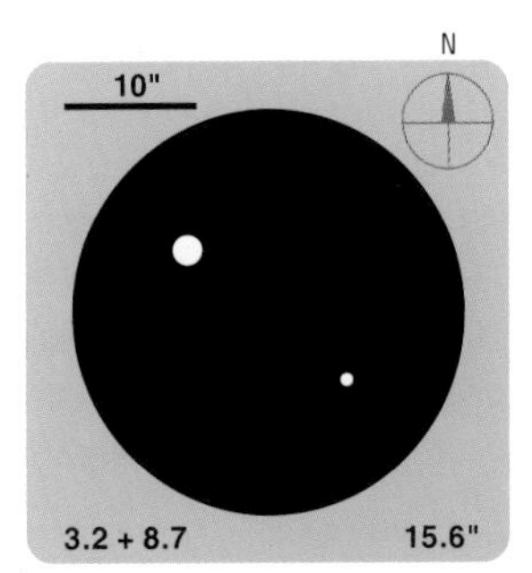

Left: Double star **Alpha Circini**, the components of which can be separated with a small telescope.

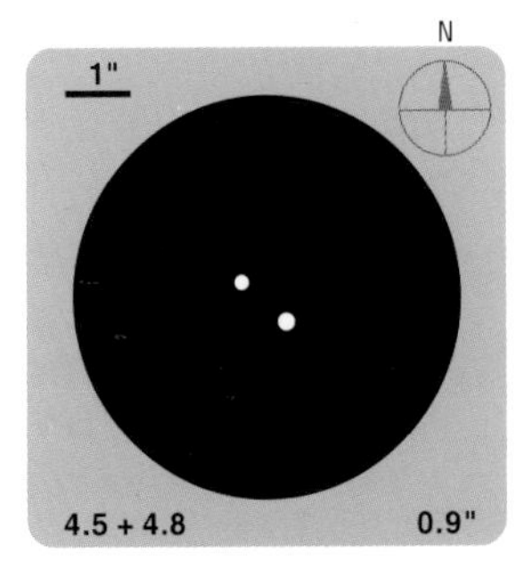

Right: **Gamma Circini** is a close visual double star, which can serve as a test object for the telescope with a 150 mm objective.

COLUMBA

Columbae Col The Dove

The southern neighbor of the Great Dog (Canis Major) and the Hare (Lepus). The constellation originated in the 17th century. It represents the dove let out by Noah from the ark when the rain had stopped, and the water started sinking. On some older maps, the constellation was called Columba Noachi, Noah's Dove. According to another myth, it was the dove which showed the Argonauts the way to the Black Sea. For this reason, it sits near the constellation of Puppis (the Stern), initially a part of the big ship Argo.

Alpha Columbae – Phakt, 2.6 mag, is a double star which can be observed with larger telescopes. It is accompanied by a faint star of 12.5 mag at an apparent distance of 13.5". According to measurements taken from the satellite HIPPARCOS and published in 1997, it lies at a distance of 268 light-years. **Beta Columbae**, 3.1 mag, at a distance of 86 light-years, is a giant star of spectral class K1. Another double star in the constellation is **Gamma Columbae**, 4.4 mag, with a faint companion of 12.7 mag at a distance 33.8" from the principal component.

Novae and Supernovae

Although astronomers report the discoveries of new stars several times a year, the real number of stars in the sky does not increase. Only the number of marks on sky maps increases that show the points where a star – not new, but rather very old – quickly flared up and disappeared. They do not mark the birth of a new star, but the explosion of an existing, inconspicuous star, the energy of which increased as much as 10,000 times. This happens in close binaries in which one component is a white dwarf. Hydrogen and helium from the other star overflow to the surface of the white dwarf until a thermonuclear reaction in the accumulated mass sets in, resulting in an explosion. After subsequent settling down, the whole process is repeated until the star explodes again (the so-called ***recurrent nova*** *–* *see p. 98**).*

More magnificent and rarer "celestial fireworks" are offered by ***supernovae****. This is not just a surface explosion on a star, which may be repeated several times, but a catastrophic explosion marking the end of the original form of the star. Only a quickly expanding gaseous envelope will remain – a* ***Supernova Remnant*** *(SNR).*

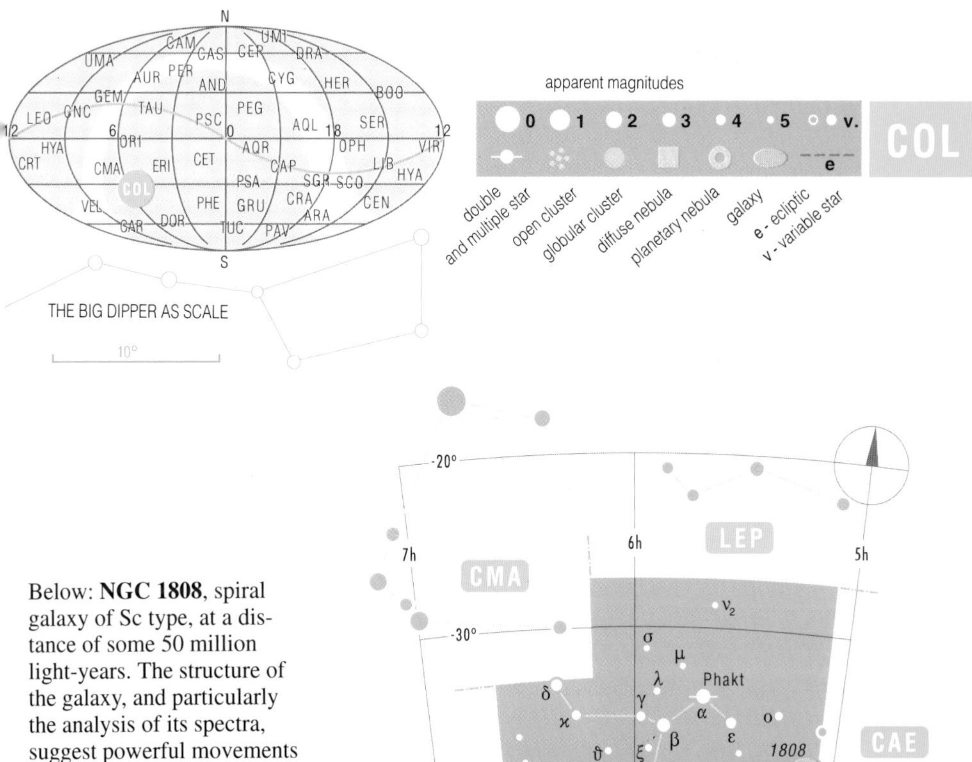

Below: **NGC 1808**, spiral galaxy of Sc type, at a distance of some 50 million light-years. The structure of the galaxy, and particularly the analysis of its spectra, suggest powerful movements in its very small nucleus, ranking the galaxy among the so-called Seyfert galaxies with extraordinarily active nuclei.

COMA BERENICES

Comae Berenices Com Berenice's Hair

The constellation recalls the real story of Queen Berenice, wife of Ptolemy III, King of Egypt (about 250 B.C.), who placed her beautiful hair on the altar as an offering for a happy return of her husband from the war. However, the hair disappeared from the temple and it was the court astronomer, Conon, who quenched the threatening scandal by "discovering" that the gods liked the offering so much that they put the hair in the sky in the form of a constellation.

A considerable part of the constellation is occupied by the extensive star cluster **Melotte 111**, one of the most beautiful open star clusters that can be observed with binoculars. The brightest star near the cluster is **Gamma Com**. The cluster contains about another 80 stars of 4 to 5 mag and fainter. It appears so big, because it is situated at a distance of only 290 light-years.

Outside the range of hobby telescopes is the cluster of galaxies in **Berenice's Hair** situated at a distance of some 300 to 400 million light-years and comprising over 1,000 galaxies. However, also a number of nearer and brighter galaxies are projected into the constellation. **NGC 4565** (below right) is one of the most typical examples of spiral galaxies observed from the side. Its distance is about 20 million light-years. It is about as big as our own Galaxy.

M 64 – NGC 4826 (the Black-Eye Galaxy) is a spiral galaxy with extraordinarily distinct dark dust clouds near the nucleus. Hence its name. Its apparent dimensions are 7.5′ × 3.5′, its distance 20 to 25 million light-years.

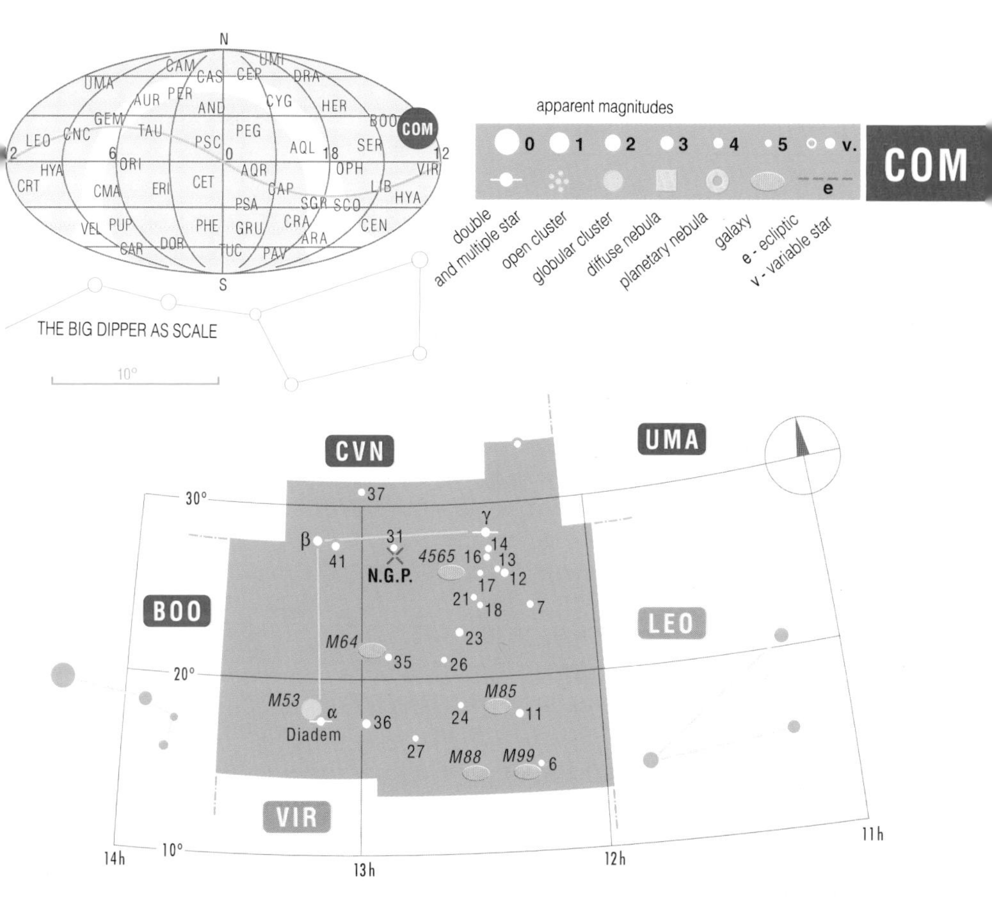

Below: Dark clouds of interstellar dust and gas can be observed along the middle plane of the galaxy **NGC 4565** which absorb the light of stars in the spiral arms of the galaxy.

Above: **N.G.P.** – North Galactic Pole. In this direction you can look perpendicularly to the galactic plane and obtain a view of the most distant parts of the universe.

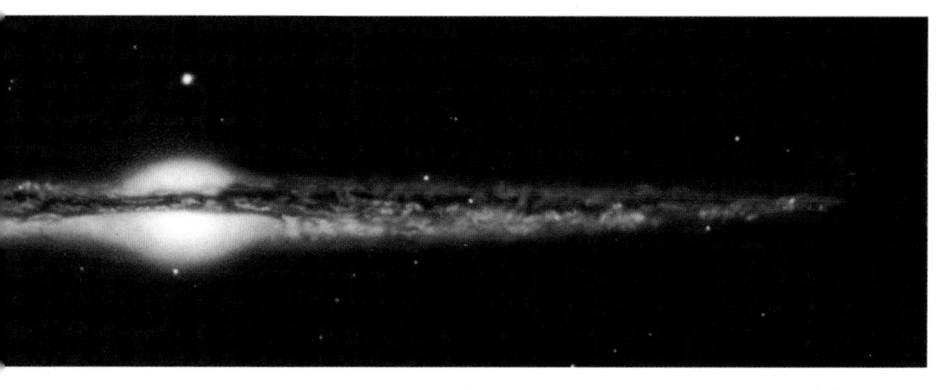

CORONA AUSTRALIS

Coronae Australis CrA The Southern Crown

South of the characteristic "teapot" decorated with the stars of the Archer (Sagittarius), you can find the counterpart of the Northern Crown (Corona Borealis) – the less conspicuous Southern Crown. Both Crowns rank among the classical constellations of Ptolemy's list dating from the 2nd century. In sky maps, the Southern Crown is usually pictured as a laurel wreath – a prerogative of gods, rulers and victors. J. Hevelius had a different idea, shown on the opposite page.

Alpha Coronae Australis, 4.1 mag, is a star of spectral class A1 at a distance of 130 light-years. **Beta Coronae Australis**, 4.1 mag, is an orange giant of spectral class K0 at a distance of some 500 light-years. The distance of 175 light-years separates us from **Delta Coronae Australis**, 4.6 mag, which is an orange giant of Class K1. **Kappa Coronae Australis** is a double star of 5.6 and 6.3 mag that can be separated with a small telescope. The mutual distance of the components is 21.4".

The globular star cluster **NGC 6541**, 6.1 mag, is big and bright. Its apparent diameter is 23'. In a small telescope it appears as a small hazy cloud. Its brightest stars of 13.0 mag can be observed with a major instrument only. The actual diameter of the star cluster is about 400 light-years. It is situated 23,000 light-years away from the Sun and 9,000 light-years from the center of our Galaxy.

Visitors to Constellations

The usual appearance of constellations is often changed by the "visitors" of celestial as well as terrestrial origin. The former include planets, asteroids, comets and rarely also new stars – novae. Some of them can be observed with the naked eye, others only with a telescope. The beginner may find it particularly difficult to determine the type of such objects. The bodies of the solar system are betrayed by their motion against the stellar background (within several hours or days), sometimes also by their appearance. The known objects can be identified by means of an astronomical yearbook and a detailed stellar atlas. If you are in doubt, do not hesitate to ask at the nearest observatory or planetarium. More than once, a nova or a comet was discovered by a hobby astronomer. However, it is more likely that chance favors those who are prepared. The number of short-term visitors to constellations includes meteors – the shining traces of particles of interplanetary matter extinguished in the atmosphere of the Earth. Fast-moving starlike objects are typical of artificial satellites of the Earth. They can be differentiated from the position lights of aircraft by binoculars and absence of noise – the satellites are noiseless.

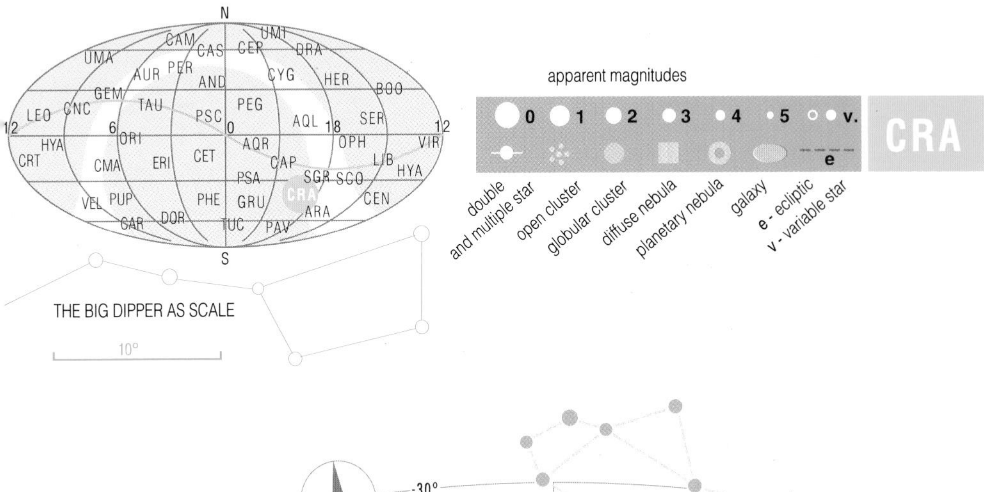

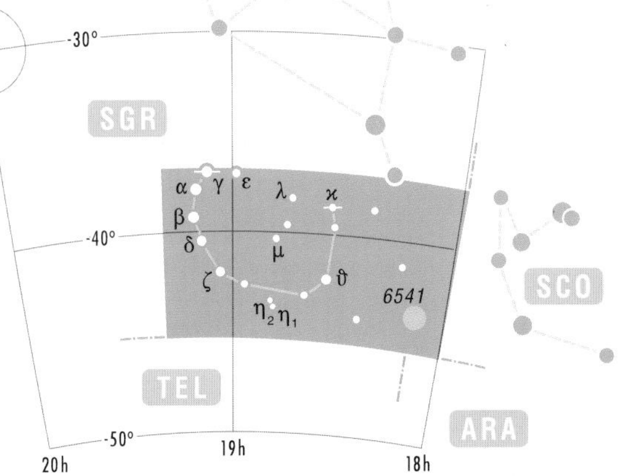

Below: **Gamma Coronae Australis** is a binary with the period of revolution of its components of 120 years. Its distance from the Earth is 58 light-years.

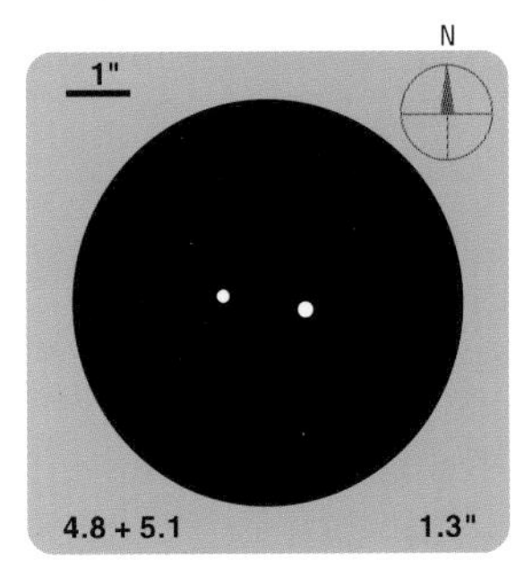

Right: The Southern Crown as pictured in Hevelius' stellar atlas of 1690.

CORONA BOREALIS

Coronae Borealis CrB The Northern Crown

One of the oldest known constellations. According to an old myth, King Minos of Crete had a daughter Ariadne who wore a jeweled crown. Dionysus (Roman, Bacchus) desired her for his wife. To convince her of his godly origin he took the princess's crown and threw it up to the sky. According to another version he did so only after Ariadne had died. The small, but distinct Northern Crown can be found easily between the Herdsman (Bootes) and Hercules.

Alpha Coronae Borealis – Gemma (the Gem) is an eclipsing binary changing its brightness between 2.2 and 2.3 mag within a period of 17.4 days. Its distance is 75 light-years. **Zeta CrB**, 4.7 mag, can be discerned with a small telescope as double star with the components of 5.1 and 6.0 mag separated by 6.3". Its distance is 470 light-years. The optical double **Nu 1,2 CrB** has two components of 5.4 and 5.6 mag at a mutual apparent distance of 370". **T Coronae Borealis**, 10.0–11.0 mag, is a star full of surprises. It is the so-called recurrent nova. In 1866 and again in 1946 its brightness increased suddenly to 2.0 mag.

The galactic cluster **ZW 7420** (Zwicky's Catalogue) contains some 400 galaxies. Its diameter in the sky is 154'. Its distance is over 1 billion light-years and it travels away from us at the velocity of some 21,000 km/s. The galactic clusters have hundreds to thousands of members mutually linked by gravitational forces.

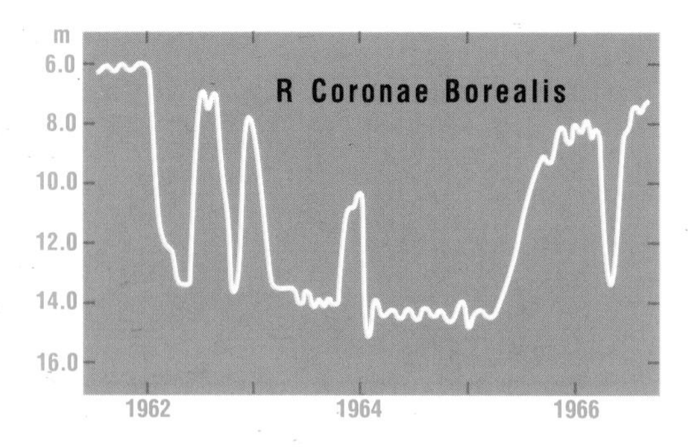

The light curve shows the changes of brightness of the irregularly variable star **R Coronae Borealis** fluctuating in an unpredictable manner within the limits of 6.0 and 15.0 mag. R CrB is a "smoker" – a star surrounded occasionally with clouds of carbon dust (soot) which absorb its light. When the clouds disperse, the star shines brightly again and attains a maximum of 6.0 mag.

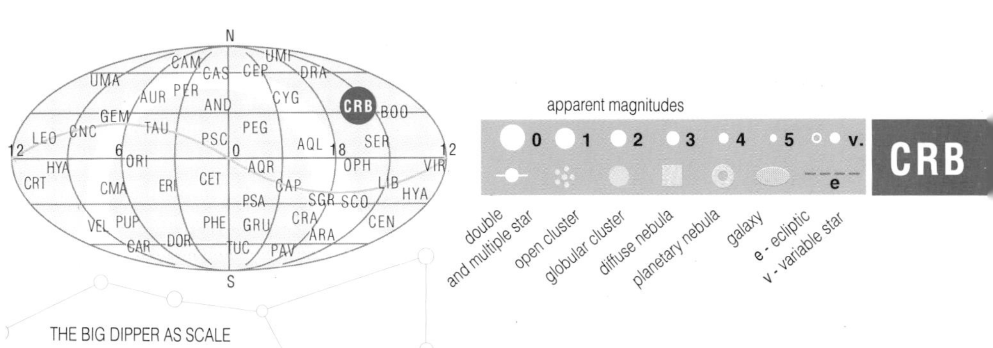

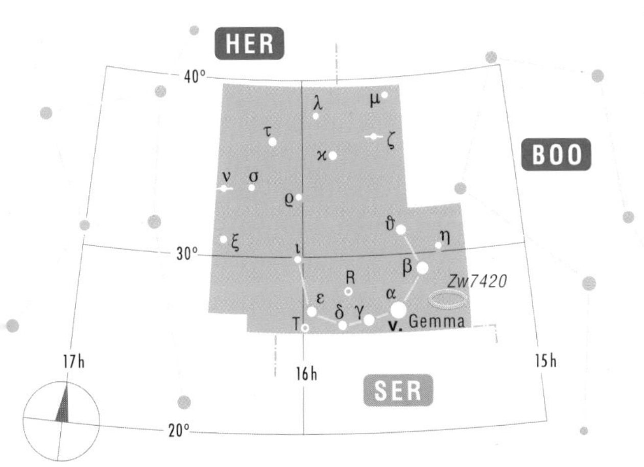

Below: Part of the galactic cluster in Corona Borealis. The small spindles and ovals are galaxies, most discs are the stars of our Galaxy between which we look into the deep space of the universe.

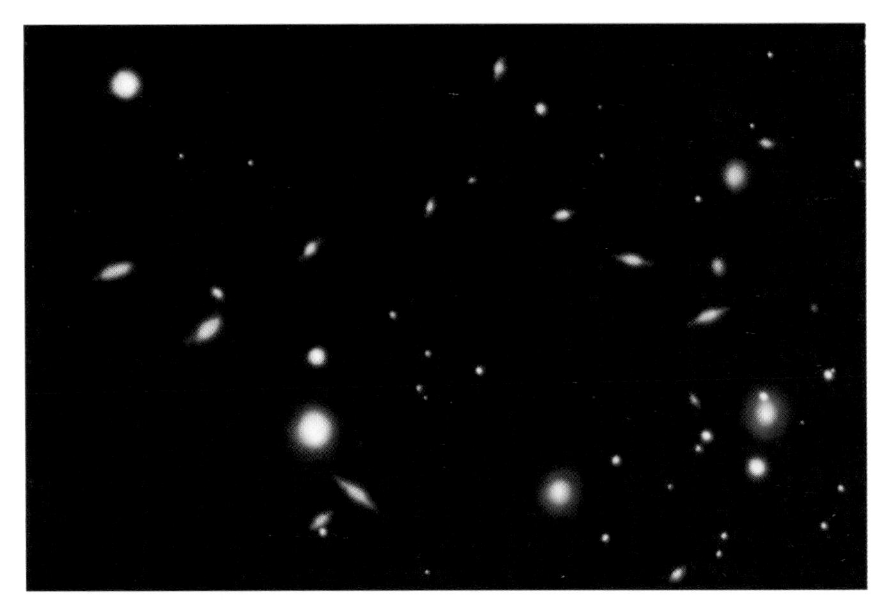

CORVUS

Corvi Crv The Crow

The conspicuous rectangle of the Crow with the adjacent Cup (Crater) can be found easily near the bright Spica in the Virgin (Virgo). Both small constellations are linked by common history (see Crater). **Alpha Corvi – Alchita** or **Al Chiba**, 4.0 mag, is a white giant at a distance of 48 light-years. **Delta Corvi – Algorab**, 3.0 mag, is a double star with a companion (9.2 mag) at an apparent distance of 24.2". The system is 88 light-years away.

The most remarkable object, although unperceivable with small telescopes, is the pair of galaxies **NGC 4038–39**, 11.0 mag (see opposite). Their unusual appearance has earned the name the **Antennae** or the **Ring-Tailed Galaxy**. Its distance is about 90 million light-years. It is one of the nearest pairs of galaxies influencing each other. Some 500 to 700 million of years ago both galaxies approached so closely to each other that a great many stars and interstellar matter were torn from their initial orbits by tital (gravitational) forces and ejected into intergalactic space in the form of long filaments. The collision has triggered massive star formation in the galaxies.

CRATER

Crateris Crt The Cup

The god Apollo sent his servant, the crow, with the cup to fetch the water of life. At the spring, the crow found a fig tree and waited until the figs were ripe. Only then did the tardy bird do as had been bidden. To explain why he was so late, he brought a water serpent (Hydra), which allegedly hindered him from taking the water. Apollo saw through his lies and put all three – the Crow, the Cup and Serpent – in the sky.

Alpha Crateris – Alkes, 4.1 mag, is a yellow giant at a distance of 174 light-years. The double star of **Gamma Crateris** has a main component of 4.1 mag and a companion of 8.9 mag, separated by 5.3". The whole system is 84 light-years from the Earth.

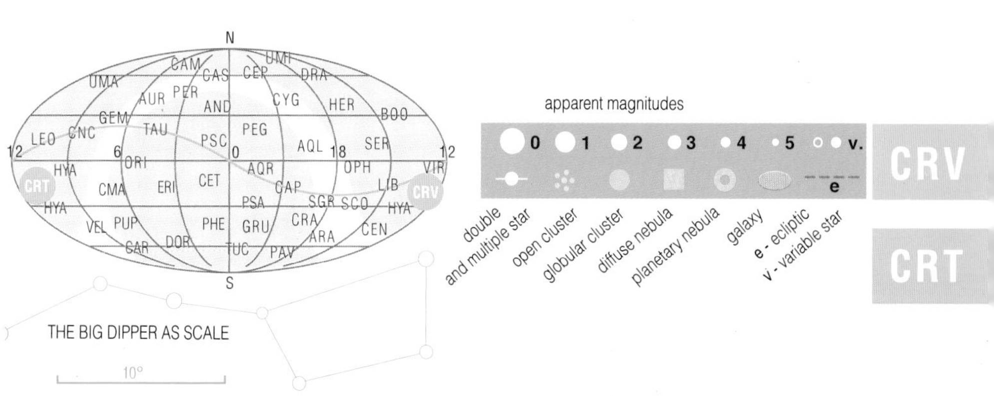

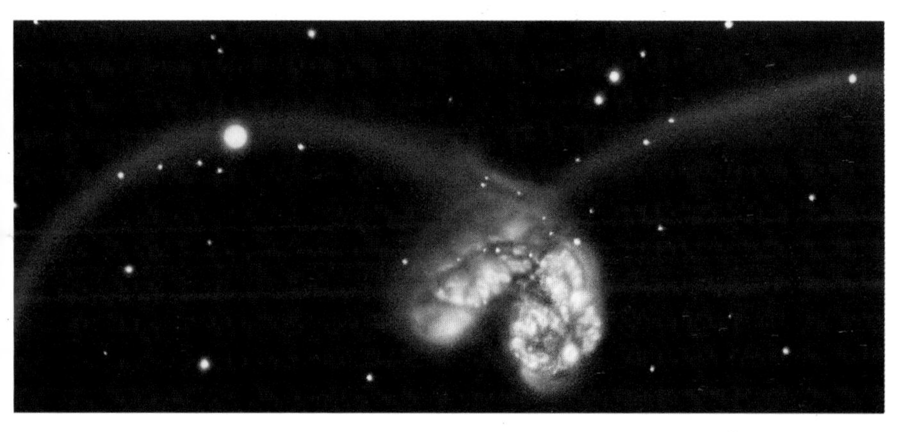

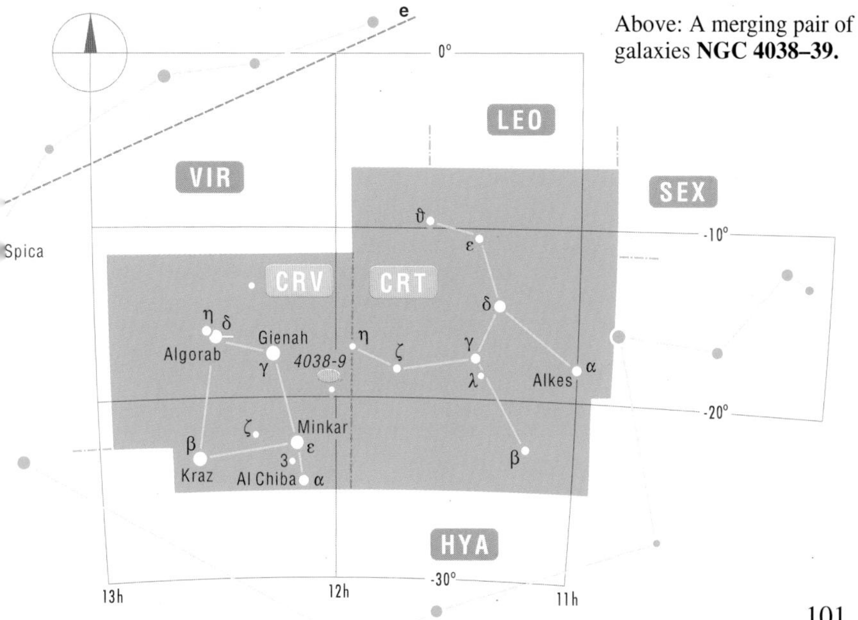

Above: A merging pair of galaxies **NGC 4038–39.**

CRUX

Crucis Cru The Southern Cross

The smallest, but the best-known and most highly admired constellation of the southern sky. It is the symbol of the southern hemisphere and is depicted on the flags of several countries. The longer arm of the Cross points at the southern celestial pole. Therefore, the Southern Cross has served as a celestial beacon to navigators for centuries. As a constellation in its own right, the Southern Cross has been recorded in sky maps since the 17th century. Earlier its stars formed part of the Centaur. The Southern Cross can be remembered together with the nearby couple of Alpha and Beta Centauri – the “pointers” to the Cross (see the map on opposite page and the picture below) and simultaneously as a security against its being mistaken for the “false cross” near the boundaries of Vela and Carina.

The brightest open star cluster **NGC 4755 – The Jewel Box** is one of the most beautiful star clusters. Within a circle of 10′ are about 50 bright stars of 6.0 to 10.0 mag. Most of them are younger blue supergiants (of several million years of age) with the luminosity of several dozen thousands of suns. The distance of the cluster is about 7,600 light-years.

The dark nebula called **The Coal Sack** appears to the naked eye as an empty area in the Milky Way, sized about 5° × 8°. It is a dark cloud of interstellar dust and gas at a distance of some 550 light-years. Its actual diameter is over 60 light-years.

Environs of the Southern Cross with **The Coal Sack** – a dark nebula.

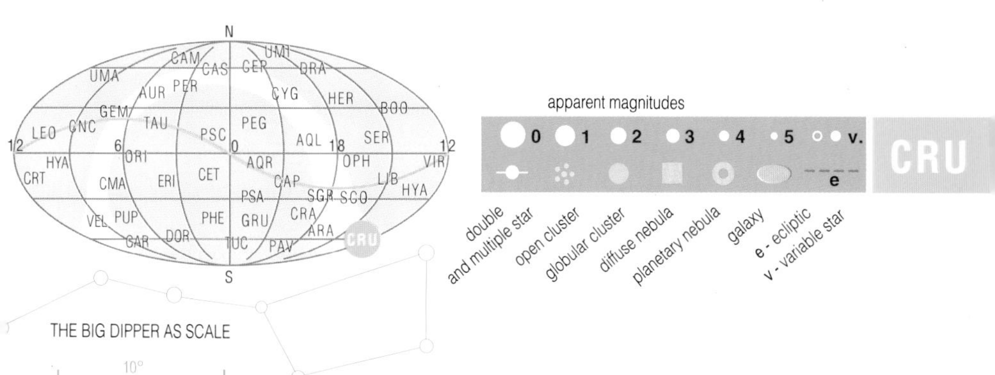

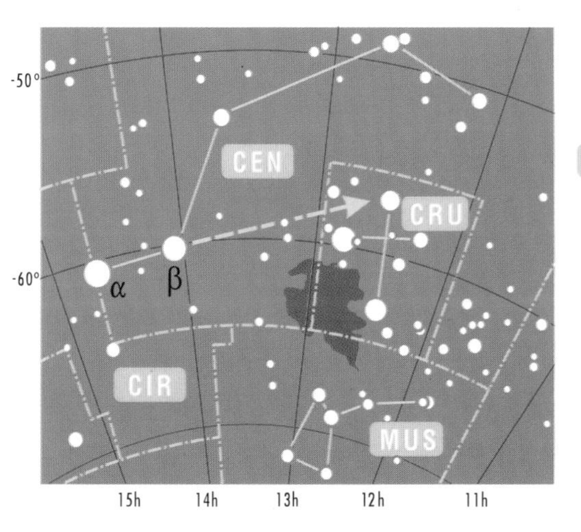

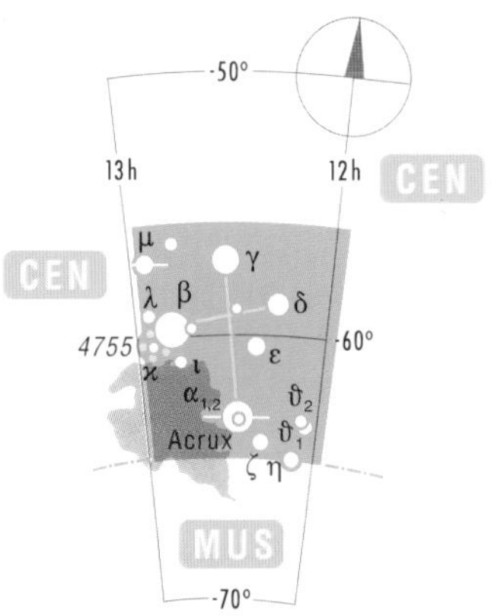

Below: Open star cluster **NGC 4755 – The Jewel Box**. A beautiful target for every telescope.

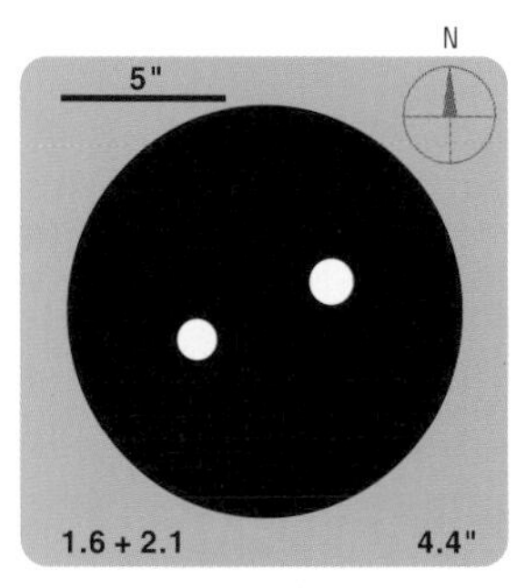

Above: **Alpha Crucis – Acrux**, 1.1 mag, one of the most beautiful double stars in the sky at a distance of about 320 light-years.

CYGNUS

Cygni Cyg The Swan

The Swan is an ancient classical constellation accompanied by a number of myths. According to one of them, it was the shape assumed by Zeus when trying to seduce the beautiful Leda of Sparta. In the sky, the Swan is a very conspicuous formation, unofficially called also the Northern Cross. It is not difficult to imagine a swan with spread wings flying along the Milky Way.

Alpha Cygni – Deneb, 1.2 mag, a bright star marking the swan "tail," is one of the corners of the "summer triangle." It is one of the biggest and brightest supergiants with a diameter of 60 suns, mass of some 25 suns and luminosity of 60,000 suns. The inconspicuous double star **61 Cygni** with components of 5.2 and 6.1 mag is of historical significance: it was the first star whose parallax was measured successfully and whose distance from the Sun was determined (F. W. Bessel, 1838). Its distance is 11.4 light-years.

Diffuse nebula **NGC 7000 – North America** (see picture below) can be observed with the naked eye or binoculars as a hazy cloud of a diameter of some 1.5° in the Milky Way. Its distance is about 2,700 light-years; its actual diameter is 100 light-years.

The nebula **NGC 7000 – North America**.

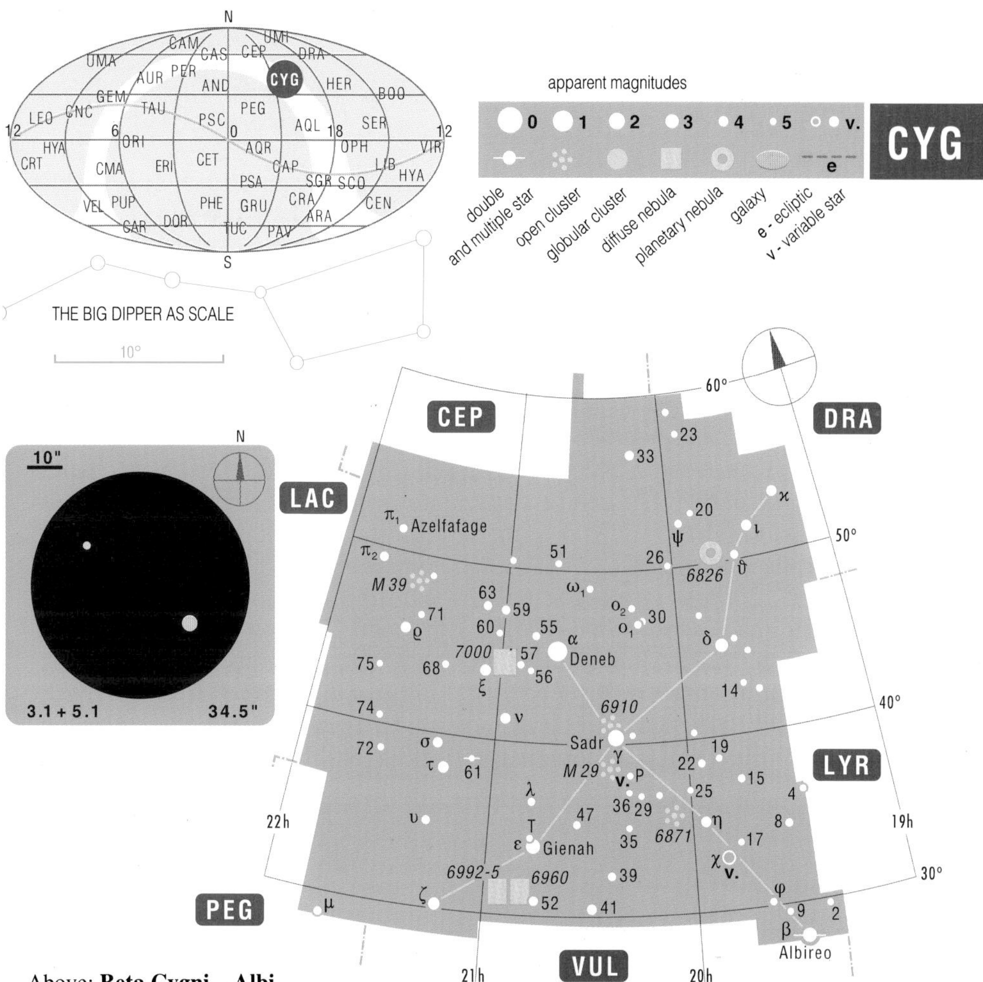

Above: **Beta Cygni – Albireo**, one of the most beautiful double stars with markedly contrasting colors. Spectral classes of its components are K3 and G0. The distance of the double is about 380 light-years.

Right: The open star cluster **M 39 – NGC 7092**, 5.0 mag, contains about 25 stars of 7.0 to 10.0 mag. Its apparent diameter is 30', its distance about 900 light-years.

DELPHINUS

Delphini Del The Dolphin

The constellation was proposed most probably by ancient navigators who knew the jumping dolphins and their friendly relation to people. One of the myths tells about the dolphin that saved the famous Greek poet Arion. According to Greek mythology, the Dolphin was placed among the constellations by Poseidon, the brother of Zeus, because it had shown him the hiding place of his desired Amphitrite.

Alpha Delphini – Sualocin, 3.8 mag, a multiple star with a companion of 13.3 mag at an apparent distance of 30". Another component of 6.0 mag is situated 9' to the southwest of the principal star. The light of the system takes some 240 years to reach the Earth. **Beta Delphini – Rotanev**, 6.6 mag, was discerned as a binary with the components of 4.0 and 4.9 mag of mutual distance merely 0.7" (at present only 0.2") by J. W. Burnham by means of a Clark's 150 mm refractor in 1873. Its distance is about 120 light-years. The mysterious names Sualocin and Rotanev begin to make sense when read back to front: Nicolaus Venator – the latinized name of the Italian observer Niccolo Cacciatore, assistant to the well-known astronomer Giuseppe Piazzi, director of the Palermo Observatory.

NGC 7006, 11.5 mag, apparent diameter 1', one of the most distant globular clusters situated more than 110,000 light-years away from the Earth.

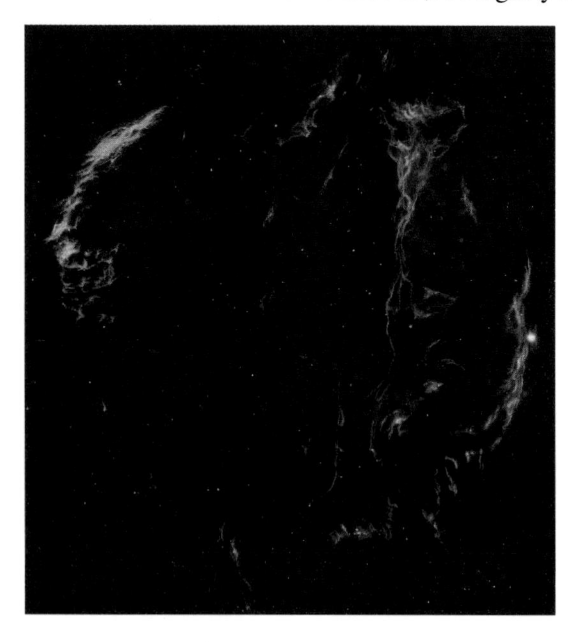

NGC 6960 and **NGC 6992–5 – The Veil Nebulae** in the Swan. The whole nebular complex has an apparent diameter of 2.6°. It is a part of the expanding "bubble" originated by the explosion of a star that flared up as a supernova at a distance of some 2,500 light-years about 50,000 years ago. (The picture belongs to p. 104 – CYGNUS.)

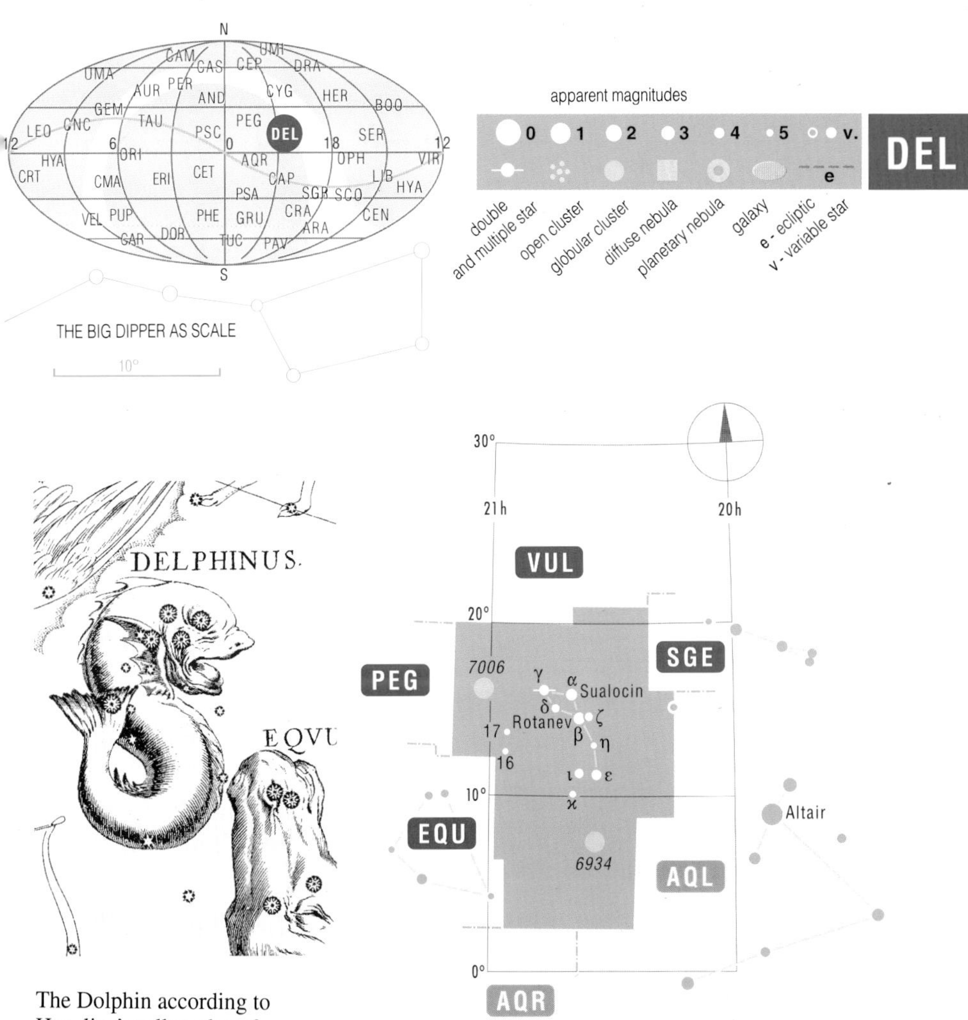

The Dolphin according to Hevelius' stellar atlas of 1690.

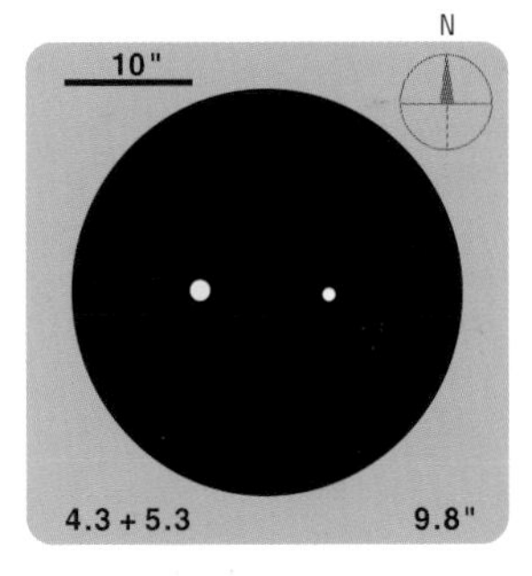

Right: **Gamma Delphini**, a beautiful double star, which can be observed with small telescopes.

DORADO

Doradus Dor The Swordfish

The constellation was proposed by the Dutch navigator P. D. Keyser in 1596 and presented by the Bavarian astronomer Johann Bayer in his stellar atlas *Uranometria* of 1603. It can be found by means of Canopus and the Large Magellanic Cloud, a substantial part of which intersects the constellation.

Beta Doradus, one of the brightest cepheids (see p. 24), varies its brightness between 3.5 and 4.1 mag in a period of 9.8 days. It is a yellow supergiant at a distance of some 1,000 light-years.

The Large Magellanic Cloud – LMC (see picture below) is a nearby, small galaxy forming, together with the Small Magellanic Cloud – SMC, satellites of our Galaxy. The diameter of LMC is about 26,000 light-years, its distance from the Earth is about 180,000 light-years. Over such – for the galactic world small – distance we can observe individual stars, star clusters and nebulae in both clouds. In February 1987, a bright supernova flared up in LMC (in the vicinity of the Tarantula Nebula) visible to the naked eye. The LMC contains a number of extraordinarily bright supergiants attaining a luminosity of as many as a million suns.

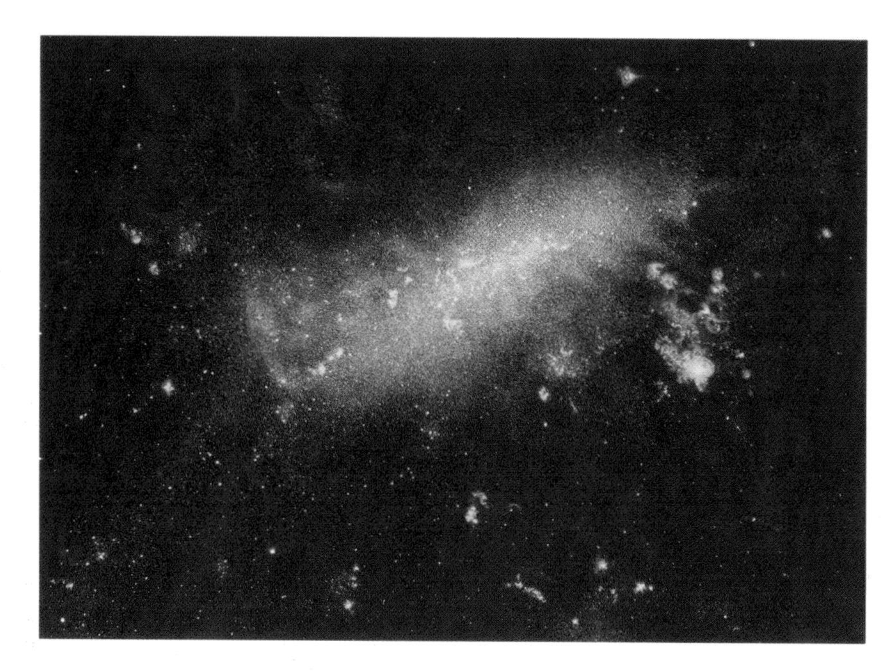

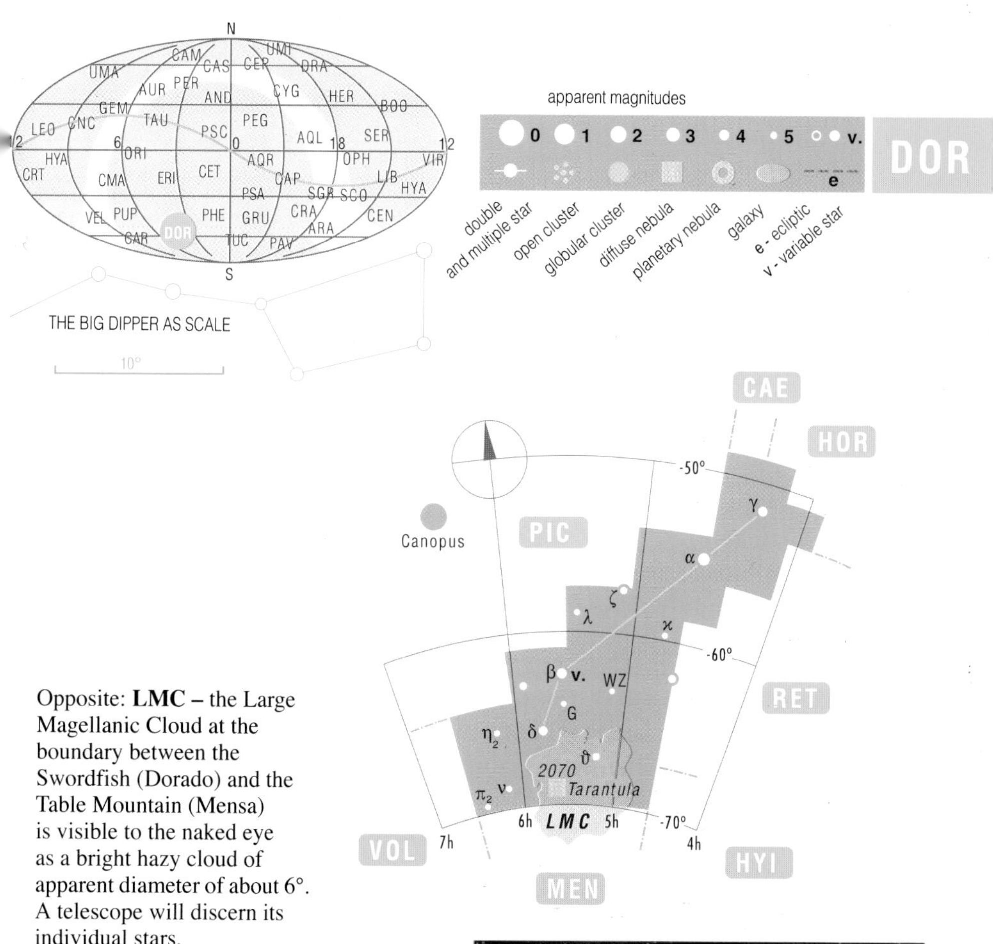

Opposite: **LMC – the Large Magellanic Cloud** at the boundary between the Swordfish (Dorado) and the Table Mountain (Mensa) is visible to the naked eye as a bright hazy cloud of apparent diameter of about 6°. A telescope will discern its individual stars.

Right: The enormous diffuse nebula **NGC 2070 – The Tarantula Nebula** is the biggest known object of its type. It can be observed with the naked eye at the edge of the LMC. Its diameter is 800 light-years with filaments extending as many as 1,700 light-years. Should it replace the well-known Great Nebula in Orion, it would cover the whole constellation and would be three times as bright as Venus.

DRACO

Draconis Dra The Dragon

According to the pictures on ancient maps, it is a snake rather than a dragon that usually has 2 to 4 legs and webbed wings. In Greek, draco meant snake. According to one myth, the dragon Ladon watched over the golden apples in the garden of the Hesperides and was killed by Hercules. In the sky the dragon's head rests at Hercules' feet not far from the bright Vega. Its snake-like body winds between both Bears.

Alpha Draconis – Thuban, 3.6 mag, is a giant star of spectral class A0 at a distance of some 310 light-years. It was the North Star of ancient Egyptians, situated at a distance of merely 10' from the celestial north pole in 2830 B.C. Due to precession, its present distance from the pole is more than 25°. **Nu Draconis – Kuma** is a pretty double star which can be observed with binoculars or a small telescope. Its white components of 4.9 and 4.9 mag are separated by 62". The distance of the double is 100 light-years. More difficult to observe is **Mu Draconis – Arrakis**, a close double star with components of 5.8 and 5.8 mag at a mutual apparent distance of only 2", situated 88 light-years away. Very faint companion of only 14.0 mag forms part of **Beta Draconis – Rastaban** which can be found in the Dragon's "head." The main star (2.8 mag) is situated at an apparent distance of 4.2" from its companion. It is a yellow supergiant some 360 light-years away.

The position of the planetary nebula **NGC 6543** (see below) is almost identical with the north pole of the ecliptic.

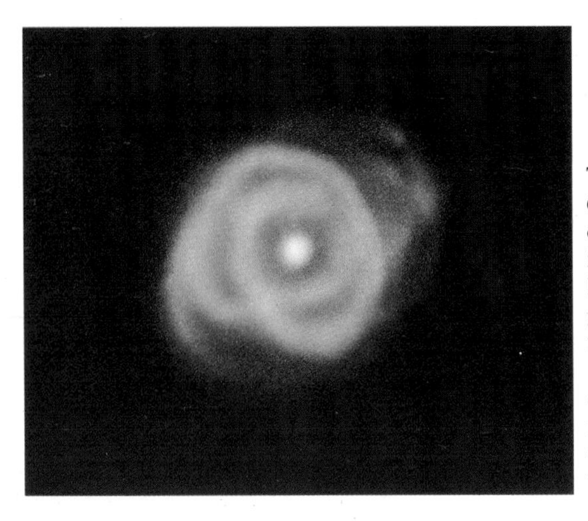

The planetary nebula **NGC 6543 – Cat's Eye** has a complex structure comprising a medley of bubbles, loops and bright nodes in gaseous clouds loosened from their central star some 1,000 years ago. We can observe it as a bright disc (9.0 mag) of apparent diameter of 20" at a distance of 2,000 to 3,000 light-years.

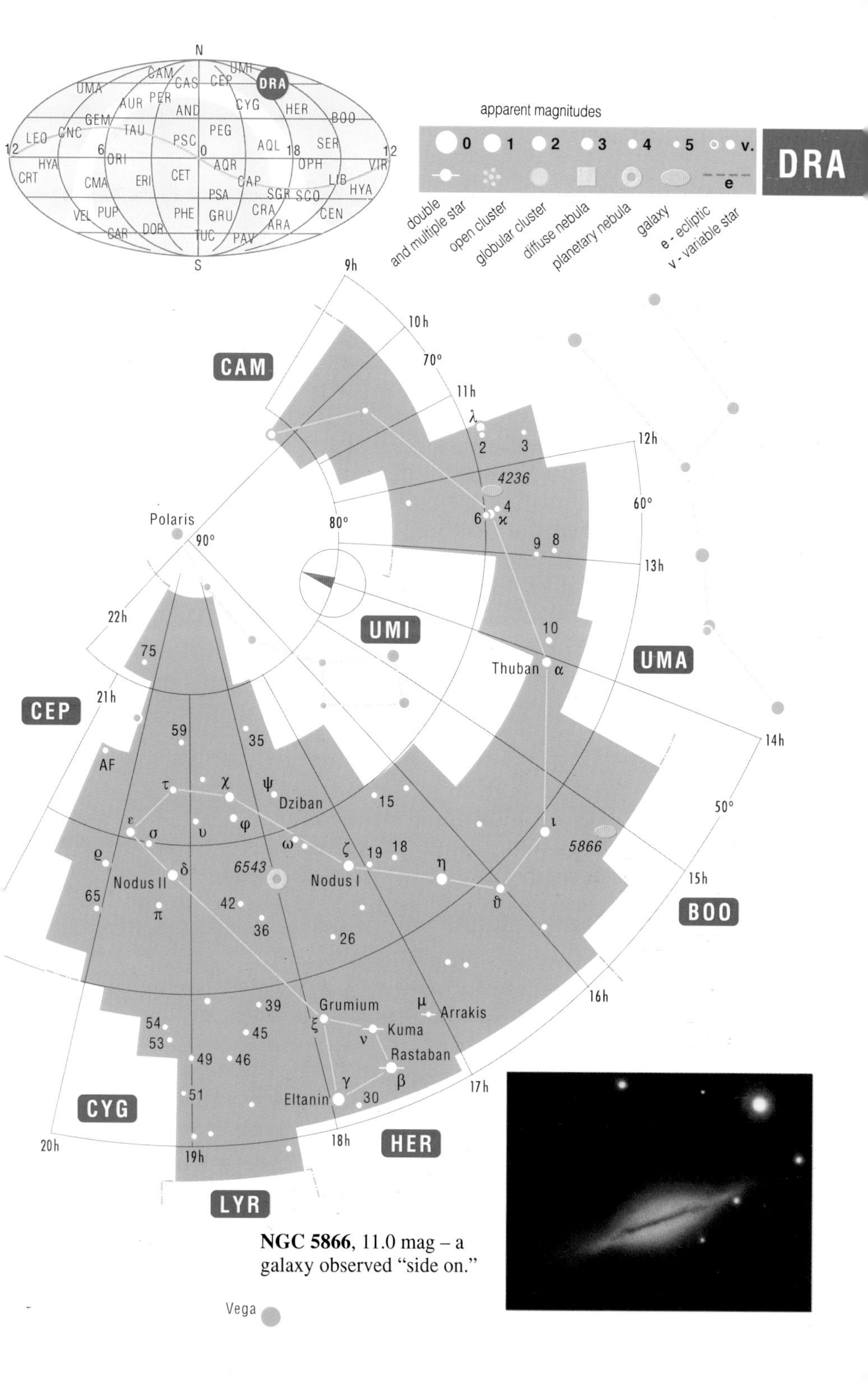

NGC 5866, 11.0 mag – a galaxy observed "side on."

EQUULEUS

Equulei Equ — The Colt, The Foal

In Greek mythology Equuleus, actually called Celeris, was the brother of the winged horse Pegasus, born from the blood of the dreadful Medusa, who was beheaded by Perseus, the liberator of Andromeda. In the sky, as in mythology, the actors of various stories meet. Equuleus was given by Hermes (Roman, Mercury) to Castor, one of the Twins (Gemini).

Equuleus ranks among the 48 classical constellations listed by Ptolemy. It is, after the Southern Cross, the second smallest constellation in the sky. In contrast to the Southern Cross, however, Equuleus is hardly discernible, as it consists of small stars of 4.0 and 5.0 mag, which can be found between the Dolphin (Delphinus) and the "head" of Pegasus.

Alpha Equulei – Kitalpha (Little Horse) is the brightest star of Equuleus, its apparent stellar magnitude being 3.9. It is a yellow giant of spectral class G0 situated at a distance of 186 light-years. **Delta Equulei**, 4.5 mag, is a star similar to the Sun at a distance of 60 light-years. **Gamma** and **Epsilon Equulei** are double stars. A bigger telescope reveals that the latter is a triple star (see the picture on the opposite page below right).

Three-Dimensional View of the Universe

If we look at the sky, our senses inform us about the directions, but not about the distances of celestial objects. Everything outside the Earth seems to be fastened to the celestial sphere at the same distance, and only the objects of our solar system show their proximity by their relatively fast motion against the background of distant stars.

Several stars near the Sun show such distinct motion that we can perceive it with our own eyes using a smaller telescope, but it would take us several years or even decades of painstaking observation.

However, we can use our imagination based on knowledge of the actual distances of objects in the universe. If we look at the constellation of Andromeda for example, we realize that the distance of Sirrah is about 100 light-years, Alamak is 3.5 times as distant, and the hazy cloud of the galaxy M 31 is 30,000 times as distant as Sirrah. It is an intellectual game, but such attempts can help us, at least in our imagination, to replace the celestial sphere with a view of the limitless space. Why not try it?

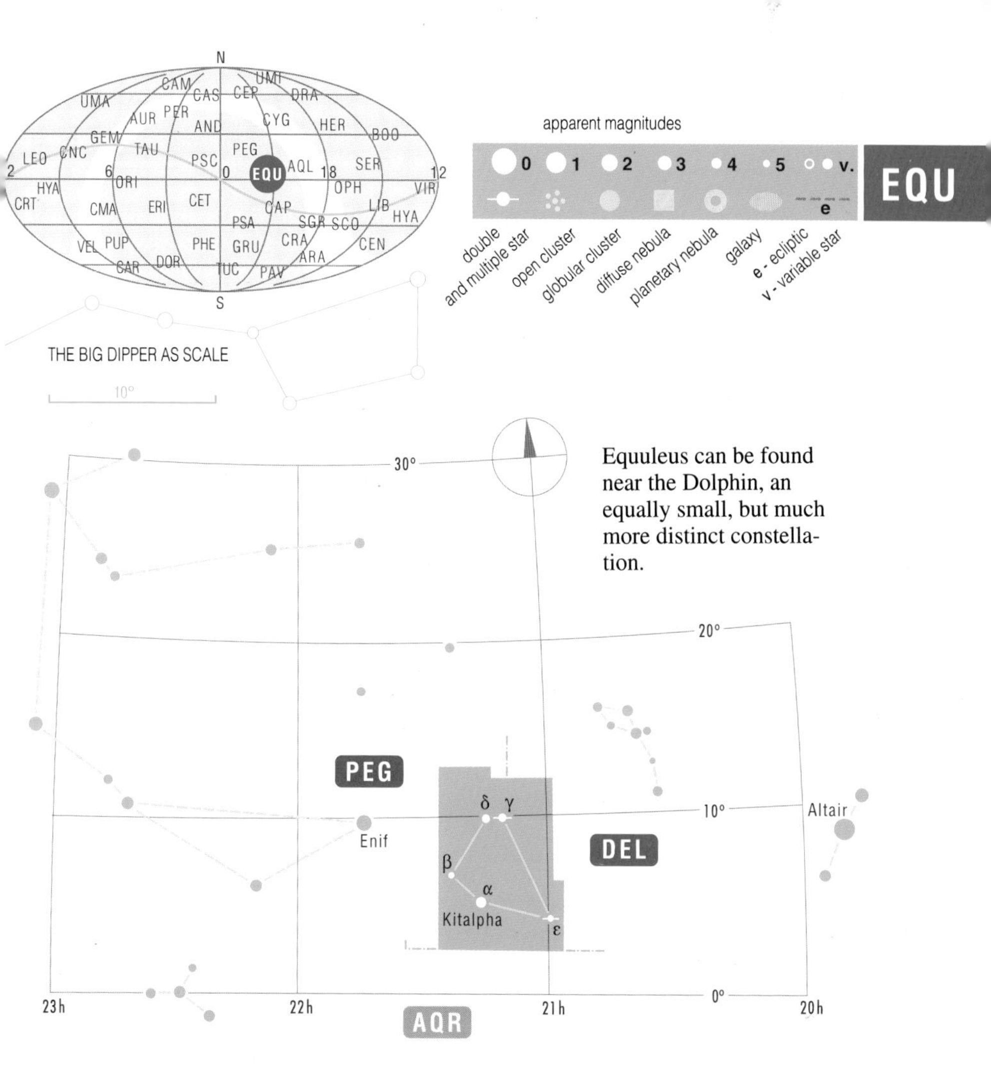

Equuleus can be found near the Dolphin, an equally small, but much more distinct constellation.

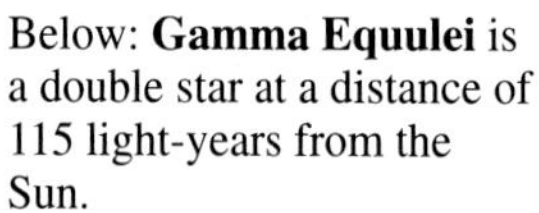

Below: **Gamma Equulei** is a double star at a distance of 115 light-years from the Sun.

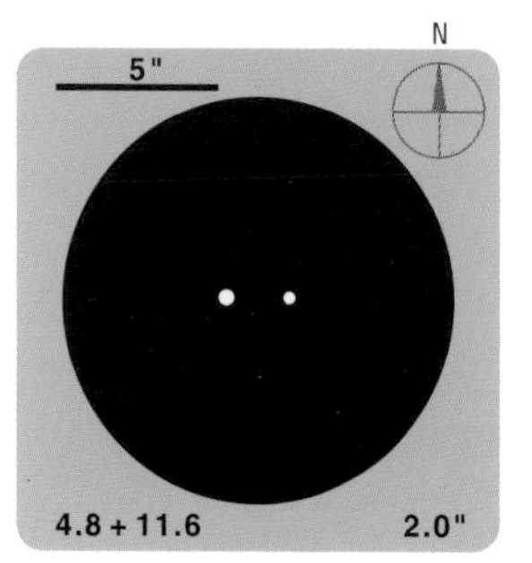

Below: Triple star **Epsilon Equulei**. The components A, B (right) have an orbital period of 101 years.

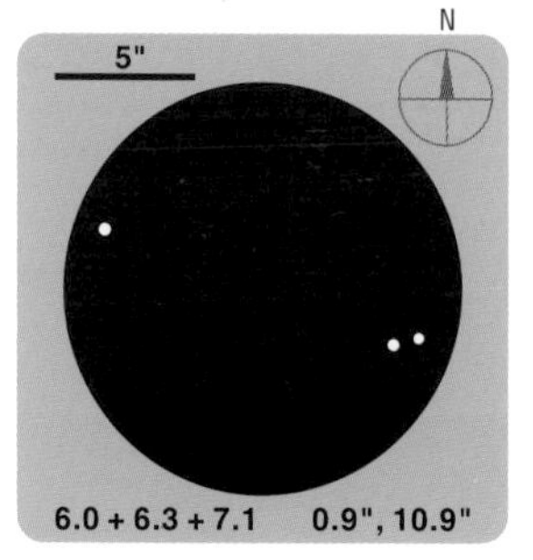

ERIDANUS

Eridani Eri The River Eridanus

The Eridanus was the river into which Phaëthon crashed after he had lost control of the runaway team of horses pulling the chariot of his father Helios, the god of the Sun. Eridanus is the sixth most extensive constellation in the sky.

Alpha Eridani – Achernar (the River's End), 0.5 mag, is a blue giant of spectral class B3 with the luminosity of 650 suns, situated at a distance of 144 light-years. **Epsilon Eridani**, 3.7 mag, ranks among the nearest neighbors of our Sun, its distance being only 10.5 light-years. With its dimensions, spectral type and luminosity like the Sun, it was one of the objects of the search for alien civilizations – which has remained fruitless. Extraordinarily interesting is the triple star **Omicron 2 Eridani – Keid**. Thanks to its short distance of 16.5 light-years, even a small telescope gives us the opportunity to see a white dwarf: its component B, the diameter of which is more than double the diameter of the Earth, but the mass of almost half the Sun. Its component C is also interesting – it is a red dwarf with a small mass equal 0.2 Sun. **Theta Eridani – Acamar** is a very pretty double star with components of 3.4 and 4.5 mag separated by 3.1".

NGC 1300 (below) – one of the best examples of a barred spiral galaxy of the SBb type.

NGC 1300 galaxy, 10.4 mag, apparent dimensions 7' × 6'.

Below: Triple star **Omicron 2 Eridani** as it appears in the telescope. Its components B and C are a white and a red dwarf respectively.

FORNAX

Fornacis For The Furnace

One of the least distinct "modern" constellations introduced by Nicolas-Louis de Lacaille in the 18th century. Its original name was Fornax Chemica – Chemical Furnace. The Furnace reaches into the meanders of Eridanus and is adjacent to the constellation of Sculptor where the south galactic pole is situated. If we look toward the Furnace, we can see one of the biggest clusters of distant galaxies.

Alpha Fornacis is a double star with components of 3.8 and 6.5 mag, separated by 4". Their orbiting period is 314 years. The distance of the system is 46 light-years.

Fornax System dE (below right) is the biggest of the seven dwarf galaxies in the Local Group. All of them are situated within the range of some 700,000 light-years and obviously are companions of our Galaxy. The Fornax System has a diameter of about 15,000 light-years.

The cluster of galaxies in Fornax is situated at the boundary between Fornax and Eridanus at a distance of 50 to 60 million light-years. It includes also **NGC 1365**, a big barred spiral galaxy. It has a diameter of more than 300,000 light-years and a mass of some 100 billion suns.

Left: Barred galaxy **NGC 1398** has a total brightness of 10.7 mag and apparent dimensions 4.5′ × 3.8′.

Below: **Fornax System dE** recalls a loose globular star cluster. Its brightest stars attain only 19.0 mag.

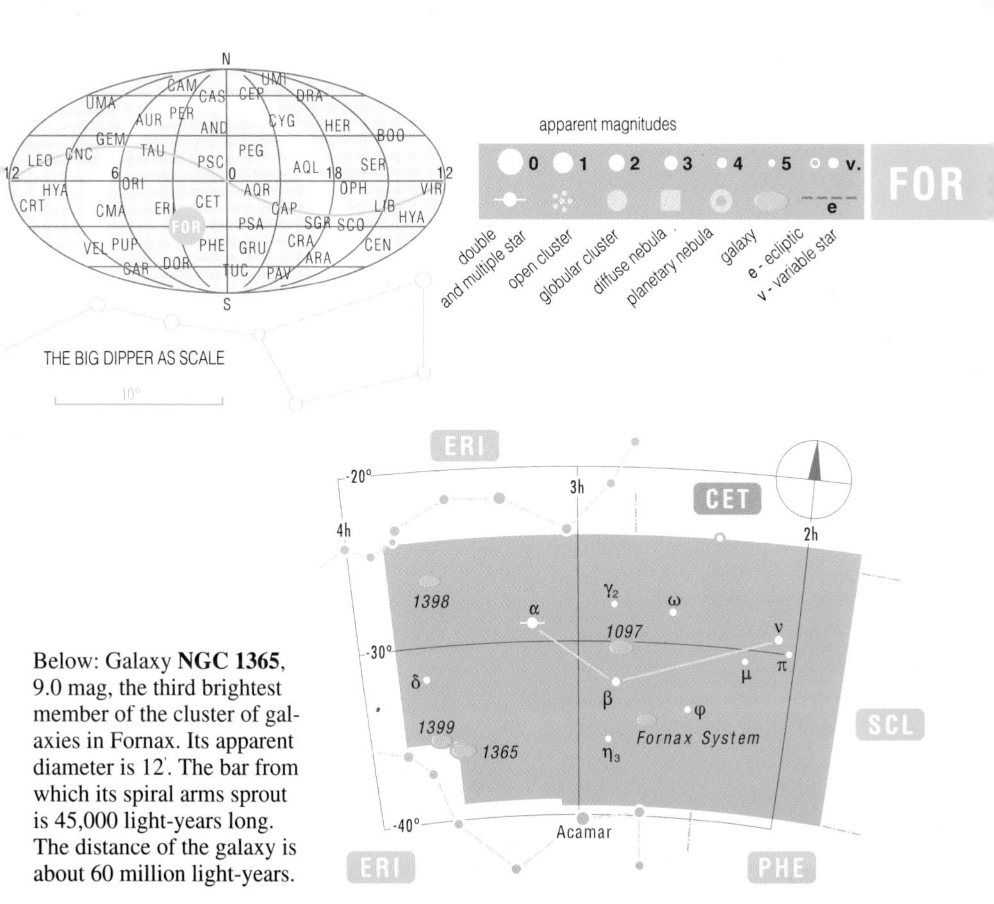

Below: Galaxy **NGC 1365**, 9.0 mag, the third brightest member of the cluster of galaxies in Fornax. Its apparent diameter is 12'. The bar from which its spiral arms sprout is 45,000 light-years long. The distance of the galaxy is about 60 million light-years.

GEMINI

Geminorum Gem The Twins

The mother of the twins Castor and Pollux was Leda, the wife of Tyndareos, King of Sparta. However, the father of Pollux was Zeus who had seduced Leda in the guise of a swan. The father of Castor, according to Greek myths, was either Zeus or Tyndareos. In the latter case Castor was mortal and Pollux immortal. After Castor's death Pollux did not want to dwell among gods alone, and so both brothers alternately travel between Olympus and Hades – the Nether World.

The brightest star in the Gemini is not Alpha, but exceptionally **Beta Geminorum – Pollux**, 1.2 mag, an orange giant of spectral class K0 at a distance of 34 light-years. **Alpha Geminorum – Castor** is a half a magnitude fainter, 1.6 mag. It is a complex sextuple system: the visible components A, B, and C are each spectroscopic binaries (betrayed by doubled lines in their spectra). Castor's distance is 52 light-years.

M 35 – NGC 2168 (see picture below), 5.5 mag, is one of the most beautiful open star clusters. Within a 0.5 circle, it is possible to observe some 200 stars, the brightest of 8.0 mag. The distance of the cluster is about 2,800 light-years.

Open star cluster **M 35 – NGC 2168**.

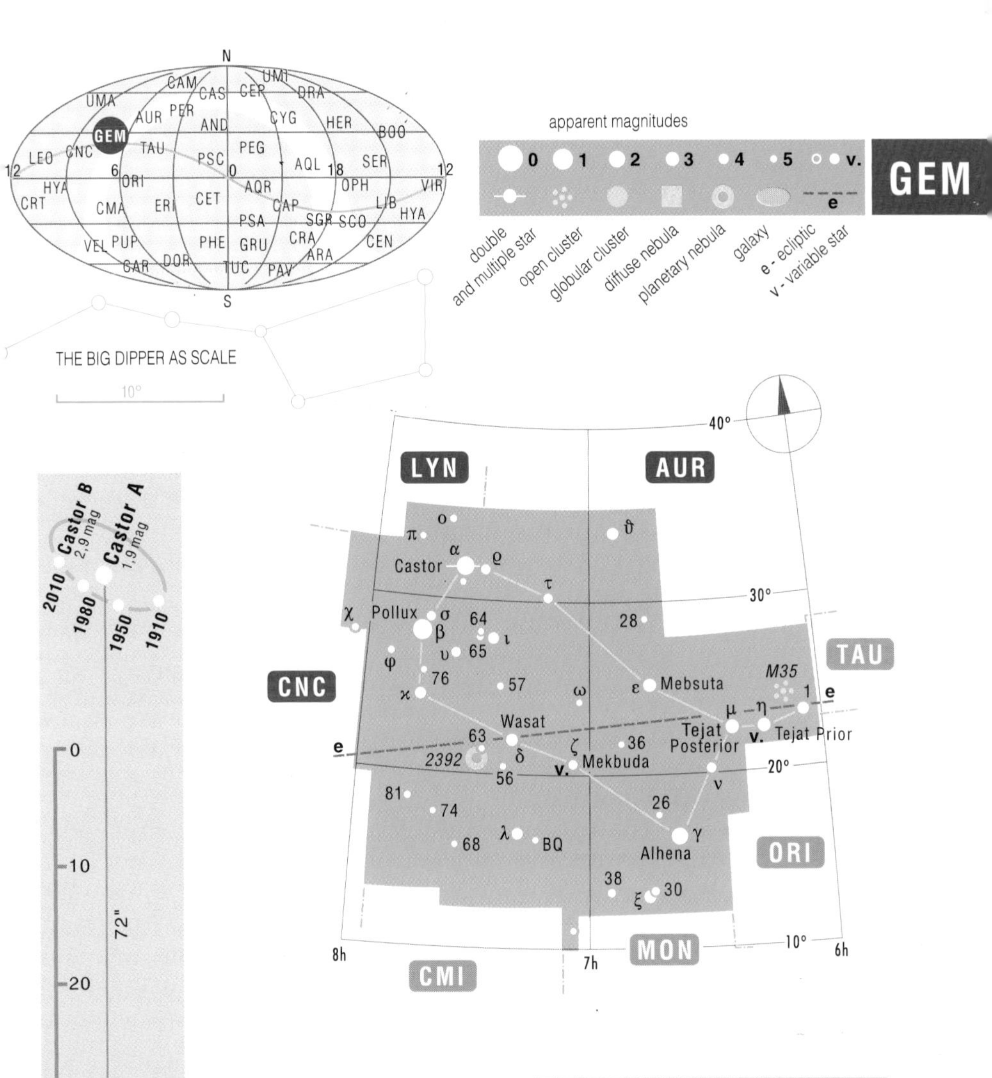

Right: **NGC 2392**, 8.9 mag, bright planetary nebula known as the **Eskimo Nebula** or the **Clown Face**. Its distance is estimated at 1,400 to 3,600 light-years.

Left: Outstanding in the multiple star **Castor** are its bright components A and B. Their apparent distance at about the year 2000 will be 4".

GRUS

Gruis Gru The Crane

The Crane was introduced to the celestial zoo by Johann Bayer in his atlas *Uranometria* in 1603. In ancient Egypt, the crane was a symbol of astronomers. The line connecting its brighter stars give the impression of a straddling bird with stretched neck and erect head. When looking for the Crane, start with the bright Fomalhaut a little to the north.

The brightest star of the Crane is **Alpha Gruis – Alnair**, 1.7 mag, a star of spectral class B7. Its luminosity is 58 times as high as that of the Sun, its diameter 3.5 suns, its mass 6 suns and its distance about 101 light-years. Alnair has a companion of 11.8 mag, at an apparent distance of 28.4". **Delta 1**, **Delta 2** and **Mu 1** and **Mu 2** are optical double stars discernible by the naked eye. Both Deltas, in turn, are double stars. Delta 1 has components of 4.0 and 12.8 mag at an apparent distance of 5.6". Delta 2 is shown on opposite page.

Planetary nebula **NGC 5148**, 14.2 mag, is a faint object of small apparent dimensions of 0.9' × 0.8' (opposite below).

Observing with Binoculars

Every hobby astronomer, from beginner to advanced, will appreciate good binoculars as a generally useful instrument. They can be used also for nature observation, are easily obtained, and are relatively inexpensive. Medium sized binoculars, of 7 × 50 magnification for example (7-fold magnification and 50 mm lens diameter) will be adequate. The binoculars can show us the craters on the Moon, the moons of Jupiter, some double stars, star clusters and nebulae. They are more than adequate for the observation of brighter comets and closer observation of constellations and fainter objects not visible with the naked eye. The field of vision of binoculars shows us a much larger part of the sky than an astronomical telescope. Before buying, you should see whether the binoculars show an acceptably sharp picture not only in the center, but also at the edges of the field of vision, whether the picture can be focused properly, and whether it is not doubled.

When you are viewing, the binoculars should always have a steady support (a tree, a window ledge, a fence, etc.) or, preferably, be mounted on a tripod to make the most of their optical quality. This applies particularly to binoculars with 10, 12 and greater magnification, which cannot be held in the hand without undesirable "shaking" of the picture.

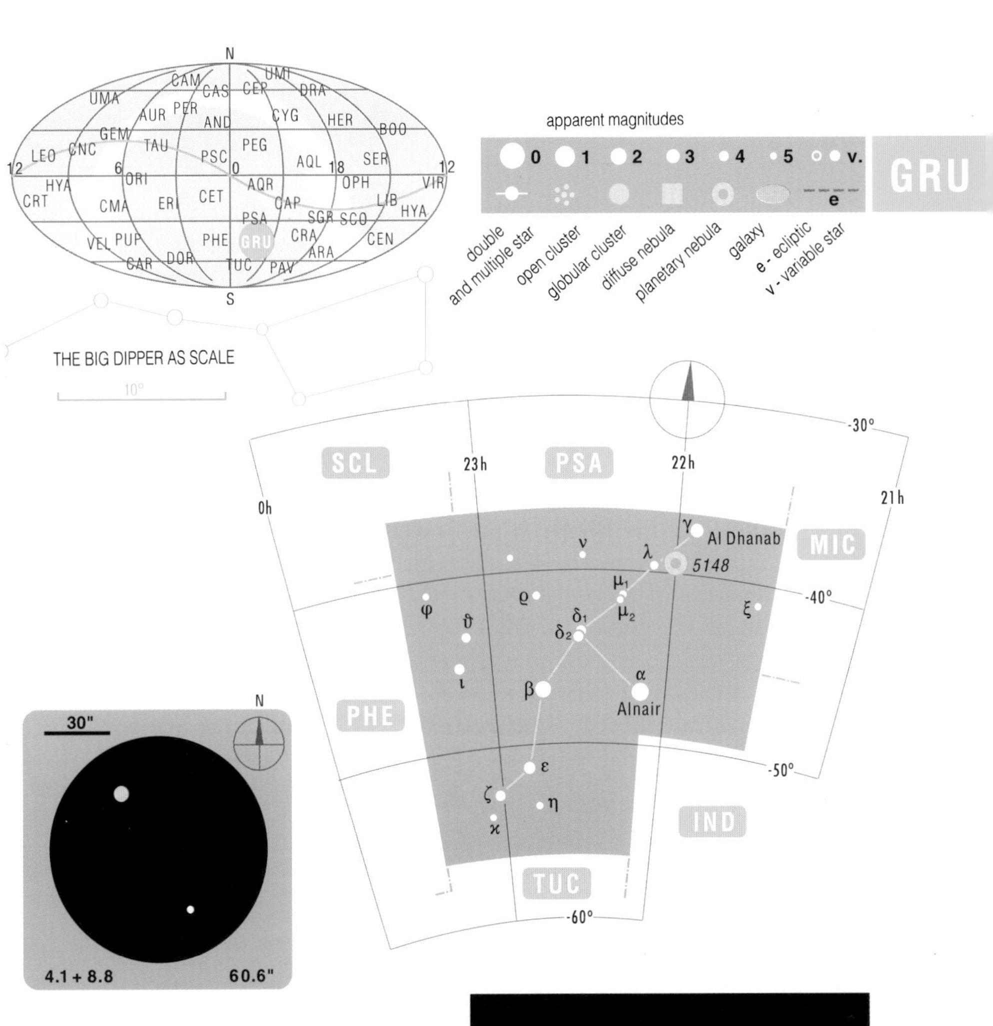

Above: Double star **Delta 2 Gruis**. Its brighter component is a star of spectral class M4 of orange hue.

Right: Planetary nebula **NGC 5148** has a very regular shape. In the picture you can see its central star and its spherical shell.

The father of the supernaturally strong Hercules was Zeus (Roman, Jupiter), his mother was Alcmene, one of Zeus' lovers. Greek mythology ascribes many heroic deeds to this archetypal "superman." Most important among them was the performance of twelve labors imposed on Hercules by King Eurystheus. In the end Zeus raised Hercules among the gods and placed him in the sky. Hercules, the fifth biggest constellation, is situated west of the bright Vega, its shape recalling the inverted letter K.

Alpha Herculis – Ras Algethi is a remarkable object: it is not only a variable star, but also a beautiful double (opposite below) and one of the biggest known stars, a red supergiant of spectral class M5. Its diameter is about 400 to 600 times that of the Sun; its luminosity is 10,000 times as high as that of the Sun. The star changes its brightness from 3.0 to 4.0 mag within a period of about 180 days. Its distance is about 380 light-years. Also **Kappa Herculis – Marfak** or **Marsic** is a pretty double star. Its components of 5.3 and 6.5 mag are separated by 30".

Globular star cluster **M 92 – NGC 6341**, 6.4 mag, apparent diameter 12', can be observed with binoculars. Its distance from the Sun is 25,000 light-years.

M 13 – NGC 6205, 5.7 mag, apparent diameter of about 20', is the brightest and most highly admired globular star cluster of the northern sky. Its actual diameter is about 350 light-years, its distance from the Sun about 23,000 light-years. It contains about a million stars. Its age is estimated at 10 billion years.

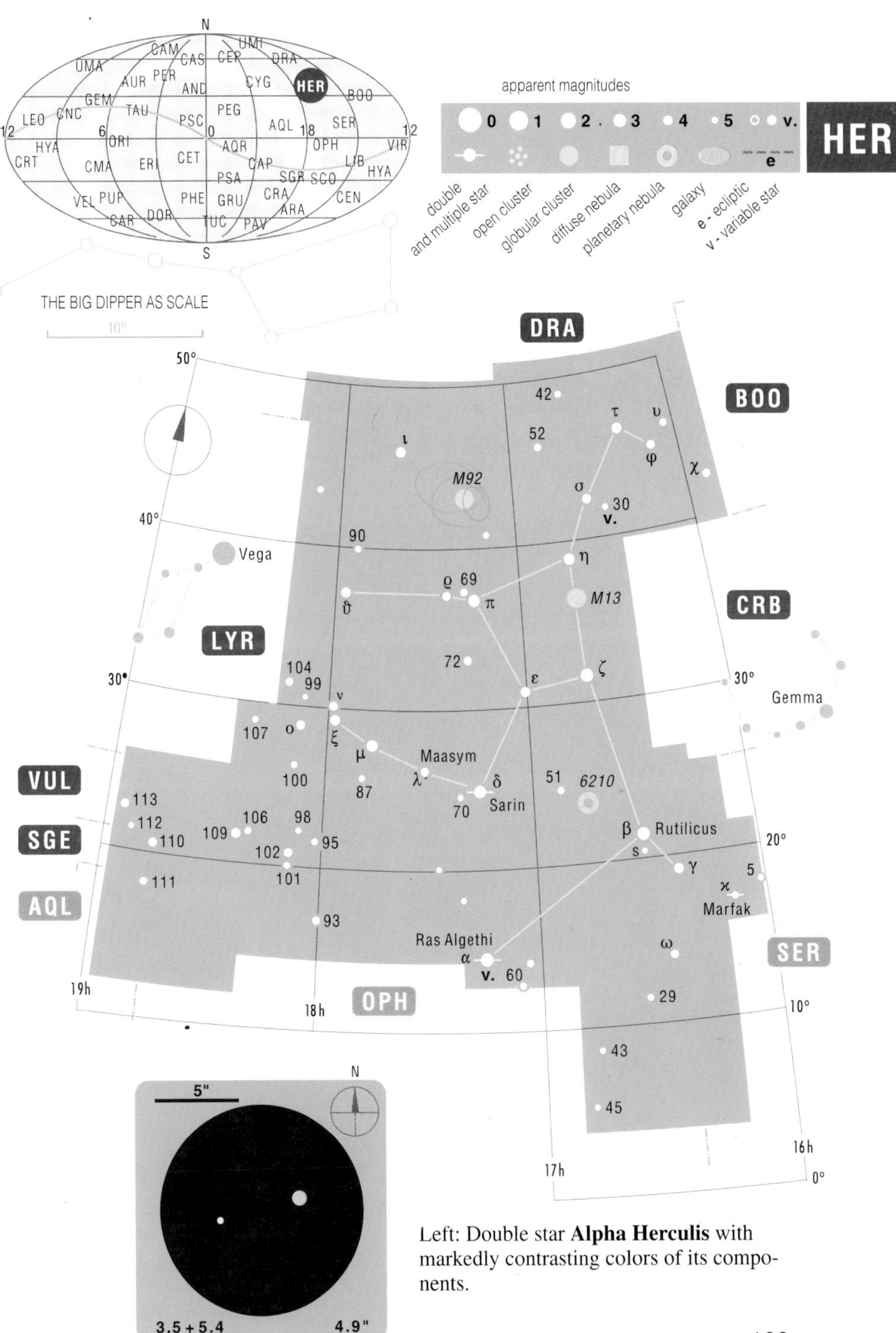

Left: Double star **Alpha Herculis** with markedly contrasting colors of its components.

HOROLOGIUM

Horologii Hor The Pendulum Clock

The stellar collection of scientific instruments established by the French astronomer Lacaille in the 18th century also includes the pendulum clock, invented by the Dutch scientist Christian Huygens in the middle of the 17th century. However, this time-keeping device of extraordinary significance was granted only a sparse group of hardly visible faint stars near the "estuary" of the River Eridanus just east of the bright Achernar.

The elongated constellation with considerably zigzagging boundaries contains only two stars brighter than 5.0 mag. The brighter of them is **Alpha Horologii**, 3.9 mag, an orange giant of spectral class K1, at a distance of 117 light-years. The long period variable star **R Horologii** changes its apparent stellar magnitude from 5.0 to 14.0 mag within a period of 13.5 months.

As shown in the orientation map on the opposite page, Horologium is situated aside of the Milky Way, in the area with an open view of the world of distant galaxies. The lower picture shows also a part of a small cluster of galaxies **ESO 249** in Horologium, situated at a distance of some 50 million light-years. The spiral galaxy on the left is very loose and more distant galaxies can be seen through its arms. The diameter of the object is about 40,000 light-years.

Cluster of galaxies in Horologium.

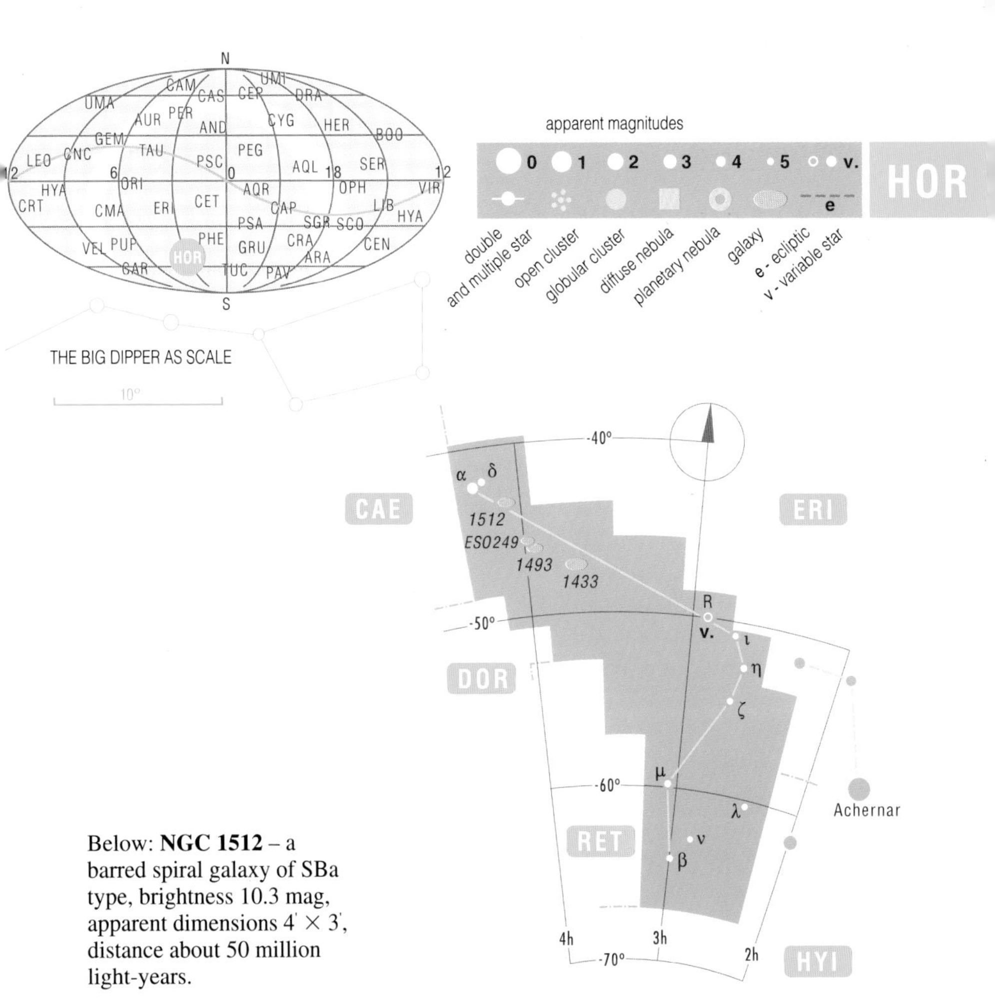

Below: **NGC 1512** – a barred spiral galaxy of SBa type, brightness 10.3 mag, apparent dimensions 4′ × 3′, distance about 50 million light-years.

HYDRA

Hydrae Hya The Water Snake

Hydra is most frequently related to the multiheaded monster or dragon fought by Hercules. The sword was no good to the hero – the beast grew two new heads for every one he had cut off. In the end Hercules conquered the monster with fire. Pictorial maps show Hydra as an enormous sea serpent. It is the largest and longest constellation: its head adjoins Cancer (the Crab) and its tail extends as far as Libra (the Scales).

Alpha Hydrae – Alphard, 2.0 mag, is the brightest star of the constellation. It is an orange giant of spectral class K3, of a diameter of 30 suns, luminosity of 230 suns, and distance of 177 light-years. **R Hydrae** is a long-period variable star with a period of 389 days. In its maximum it attains 4.0 mag, in its minimum 10.0 mag.

M 83 – NGC 5236 (see overleaf on p. 128) – one of the brightest and most beautiful galaxies of the southern sky. It appears as a hazy cloud of 7.5 mag and apparent dimensions of 11′ × 10′. Its actual diameter is about 35,000 light-years, its distance about 12 million light-years.

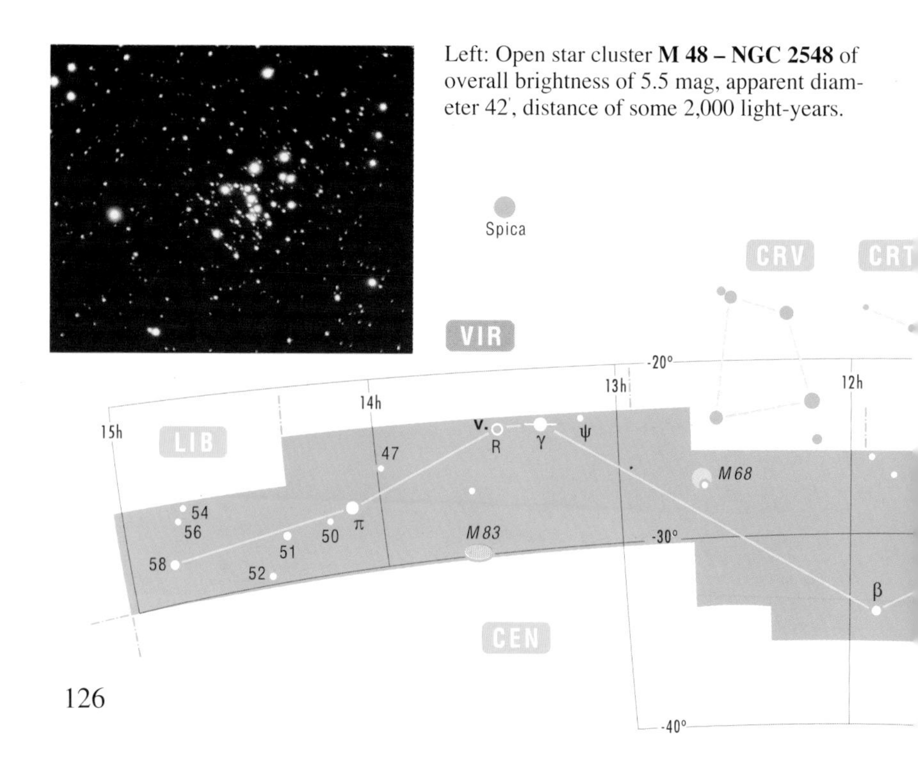

Left: Open star cluster **M 48 – NGC 2548** of overall brightness of 5.5 mag, apparent diameter 42′, distance of some 2,000 light-years.

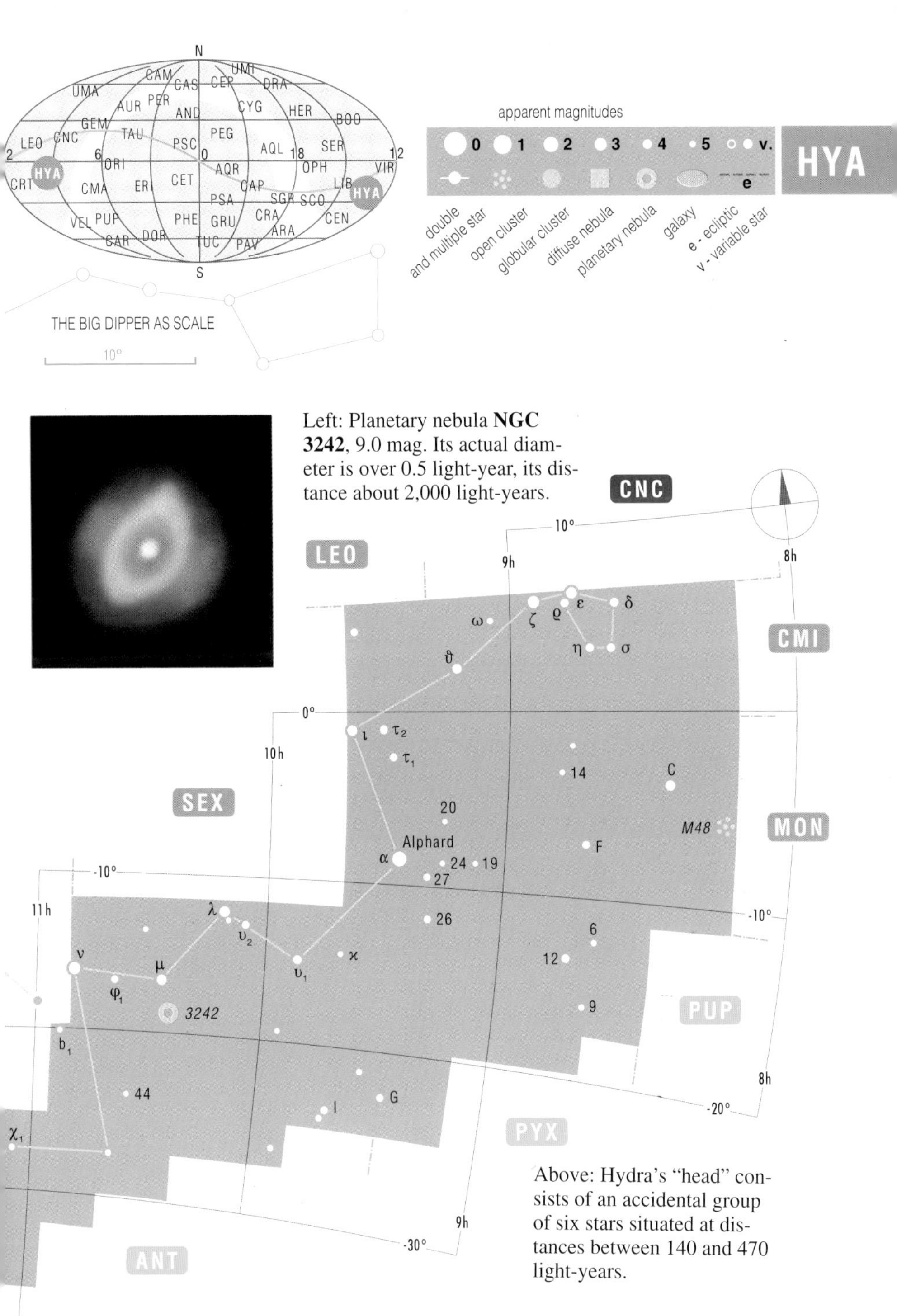

Left: Planetary nebula **NGC 3242**, 9.0 mag. Its actual diameter is over 0.5 light-year, its distance about 2,000 light-years.

Above: Hydra's "head" consists of an accidental group of six stars situated at distances between 140 and 470 light-years.

HYDRUS

Hydri Hyi The Lesser Water Snake

The origin of this and 11 other constellations is due to Dutch navigators Pieter Dirckszoon Keyser and Frederick de Houtman who proposed the supplements to the maps of the southern stellar sky at the end of the 16th century. The names of their constellations were adopted and "codified" by Johann Bayer in his atlas *Uranometria*. Consequently, Hydrus is a modern constellation not enveiled in myths as its ancient counterpart, Hydra.

Its brightest star is **Beta Hydri**, 2.8 mag. Its diameter is twice the diameter of the Sun, its mass 1.4 sun, its surface temperature 6,300 K and its distance 24.4 light-years. Only slightly fainter (2.9 mag) is **Alpha Hydri** situated at a distance of 71 light-years. The most distant in the triangular figure of the Lesser Water Snake is **Gamma Hydri**, 3.2 mag, the light of which reaches the Earth in 214 years. It is a red giant of spectral class M2.

Galaxy **M 83 – NGC 5236** in Hydra. (The picture belongs to p. 126.)

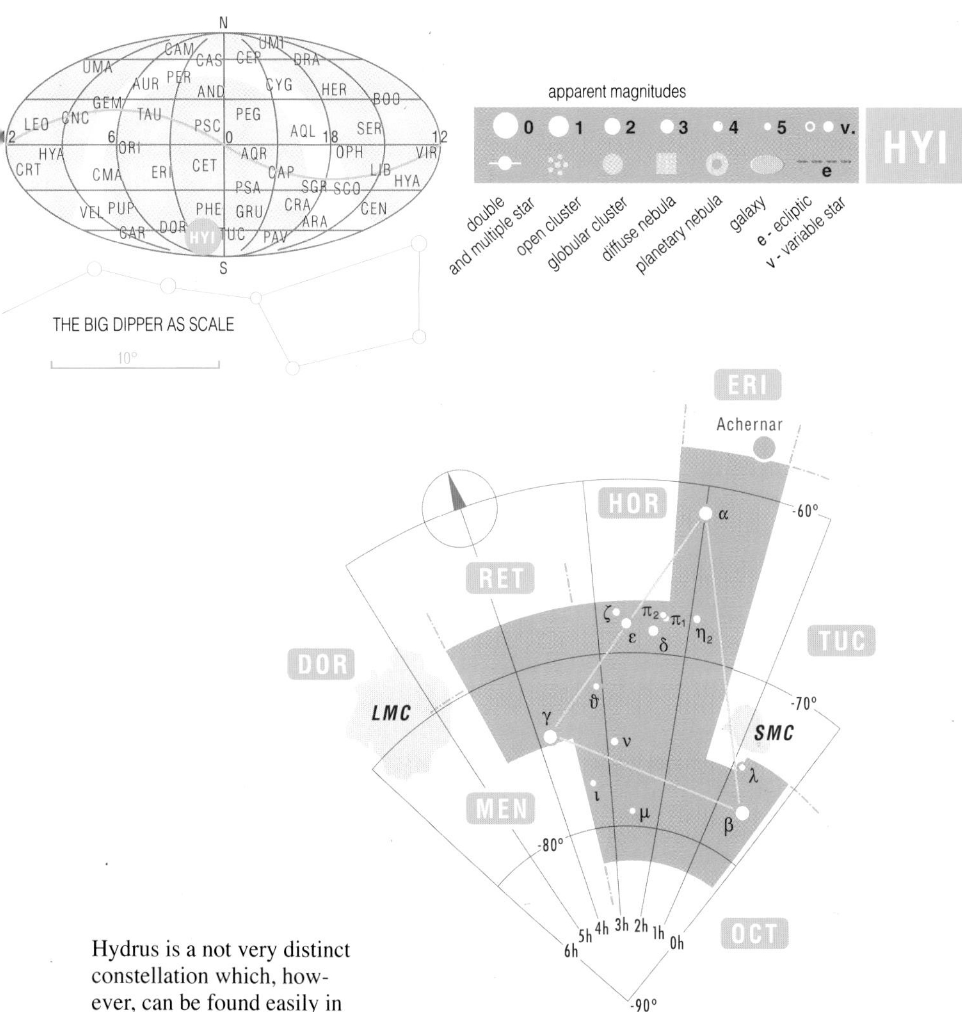

Hydrus is a not very distinct constellation which, however, can be found easily in the triangle defined by both Magellanic Clouds and the bright Achernar of Eridanus.

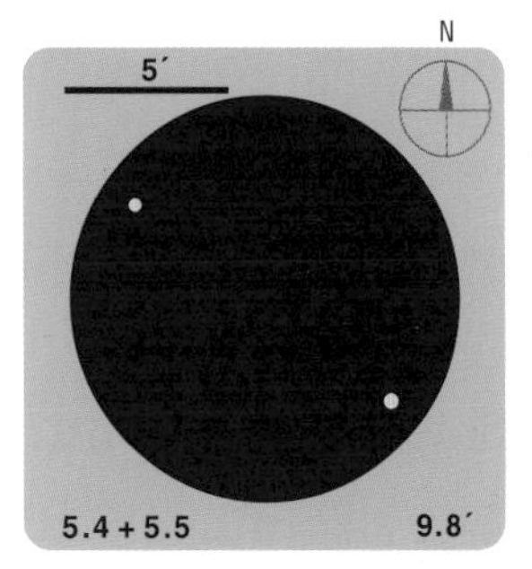

The pair of stars **Pi 1** and **Pi 2 Hydri** can be discerned with the naked eye, if permitted by observation conditions, because both stars are very faint.

INDUS

Indi Ind The Indian

The constellation was proposed by P. D. Keyser, the Dutch navigator, in honor of the natives of America at the end of the 16th century. It was introduced to stellar maps by J. Bayer in 1603. As in the case of many other modern constellations, it is hardly possible to shape the sparsely dispersed stars into the figure of a North American native. The constellation merges with its environs. Its presence is betrayed by three birds: the Crane, the Peacock and the Toucan.

Theta Indi is a double star which can be observed with a small telescope. Its components of 4.5 and 6.9 mag are separated by 6.5". Its distance is about 97 light-years. The star **Epsilon Indi**, 4.7 mag, at a distance of merely 11.8 light-years from the Sun (opposite below), is remarkable. It travels quickly, shifting 4.7" per year. In 2640 it will pass to the adjacent constellation of Toucan.

IC 5152 is an irregular galaxy at a distance of less than 15 million light-years, enabling the separation of its brightest stars. In the foreground you can see a bright star of 8.0 mag.

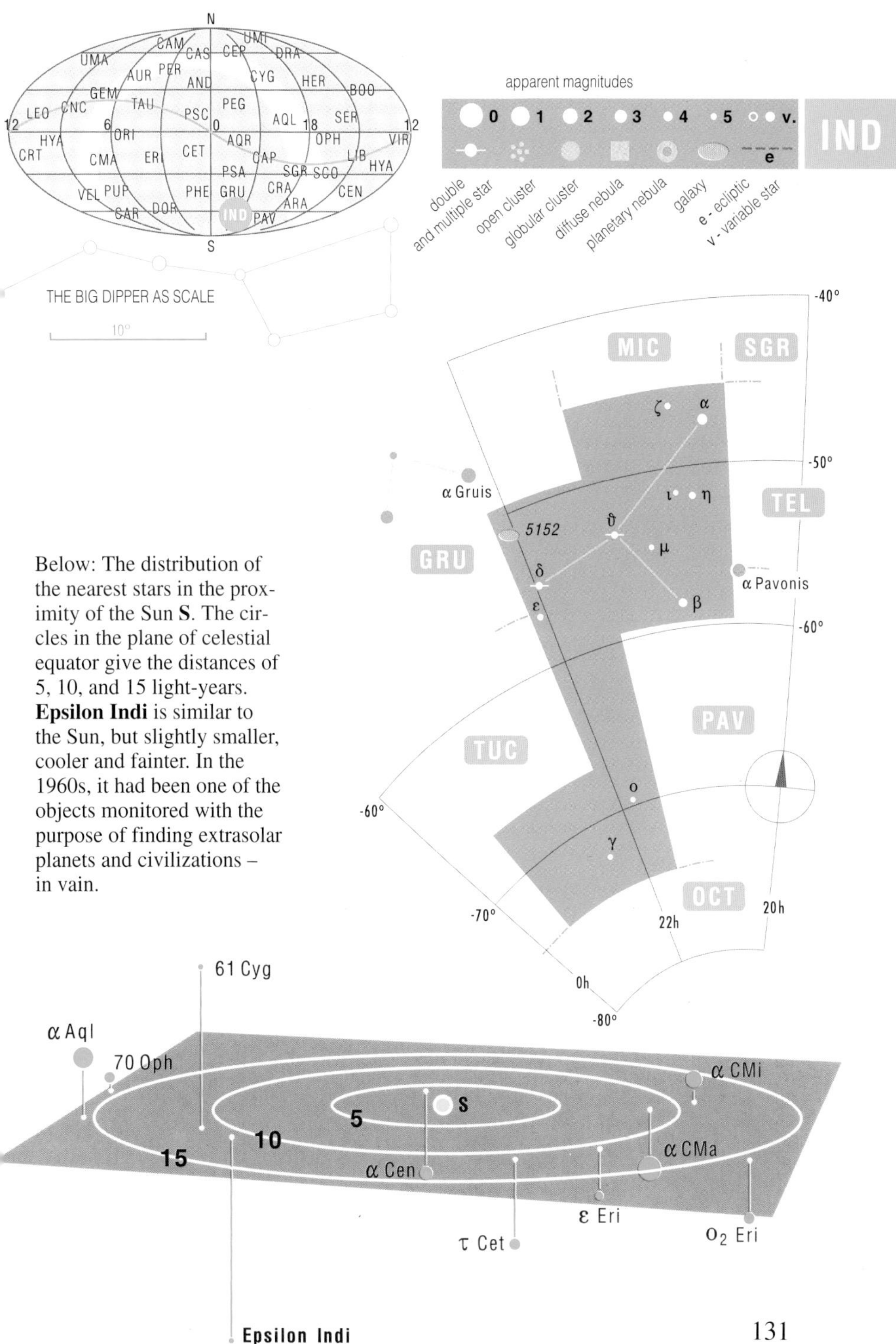

Below: The distribution of the nearest stars in the proximity of the Sun **S**. The circles in the plane of celestial equator give the distances of 5, 10, and 15 light-years. **Epsilon Indi** is similar to the Sun, but slightly smaller, cooler and fainter. In the 1960s, it had been one of the objects monitored with the purpose of finding extrasolar planets and civilizations – in vain.

LACERTA

Lacertae Lac The Lizard

The zigzagging row of faint stars in the Milky Way between Cassiopeia and the Swan inspired Johann Hevelius, astronomer from Danzig, at the end of the 17th century to form the constellation of the Lizard. His idea, reproduced from his atlas of 1690, is illustrated on the opposite page. The Lizard can be found with the assistance of the group of five stars (2–5–4–Alpha–Beta Lacertae), forming the letter W like a "diminutive Cassiopeia."

The constellation has only one star brighter than 4.0 mag, **Alpha Lacertae**, 3.8 mag, at a distance of 102 light-years. Fainter, but more interesting is **4 Lacertae**, 4.6 mag, a white supergiant from which we are separated by more than 1,500 light-years. Considerable attention of astronomers is attracted by **BL Lacertae**, initially believed to be a variable star fluctuating between 13.0 and 16.0 mag. Actually it is the nucleus of a very distant galaxy with an exceptionally strong source of energy and variable brightness. The open star cluster **NGC 7209**, total brightness 7.7 mag, contains some 100 stars of 10.0 mag and fainter, dispersed within 20′.

Messier's Objects

Charles Messier (1730–1817) was a French astronomer, excellent observer and connoisseur of stellar sky, and the discoverer of 15 comets. In 1758, he discovered a nebula in the constellation of Taurus (the Bull), which obtained the consecutive number 1 in his catalogue (M1 – the well-known Crab Nebula). To eliminate the possibility of confusion with comets, Messier began the systematic compilation of a catalogue of star clusters, nebulae and galaxies based on his own observations and those of other discoverers. In 1783 Messier's catalogue was published; it comprised 103 objects, about one third of which were his own discoveries. Later on, objects M 104 to M 110 were added on the basis of Messier's unpublished notes.

Messier's objects are very popular and form the object of special literature including pictorial atlases. "Messier's marathons" are very popular among the friends of astronomy; their purpose is to find as many objects as possible in the course of one observation night. However, Messier's objects do not represent, by far, all deep-sky objects that can be observed with smaller telescopes. They do not include, for example, the magnificent double star cluster "chi and h" in Perseus or the numerous jewels of the southern sky.

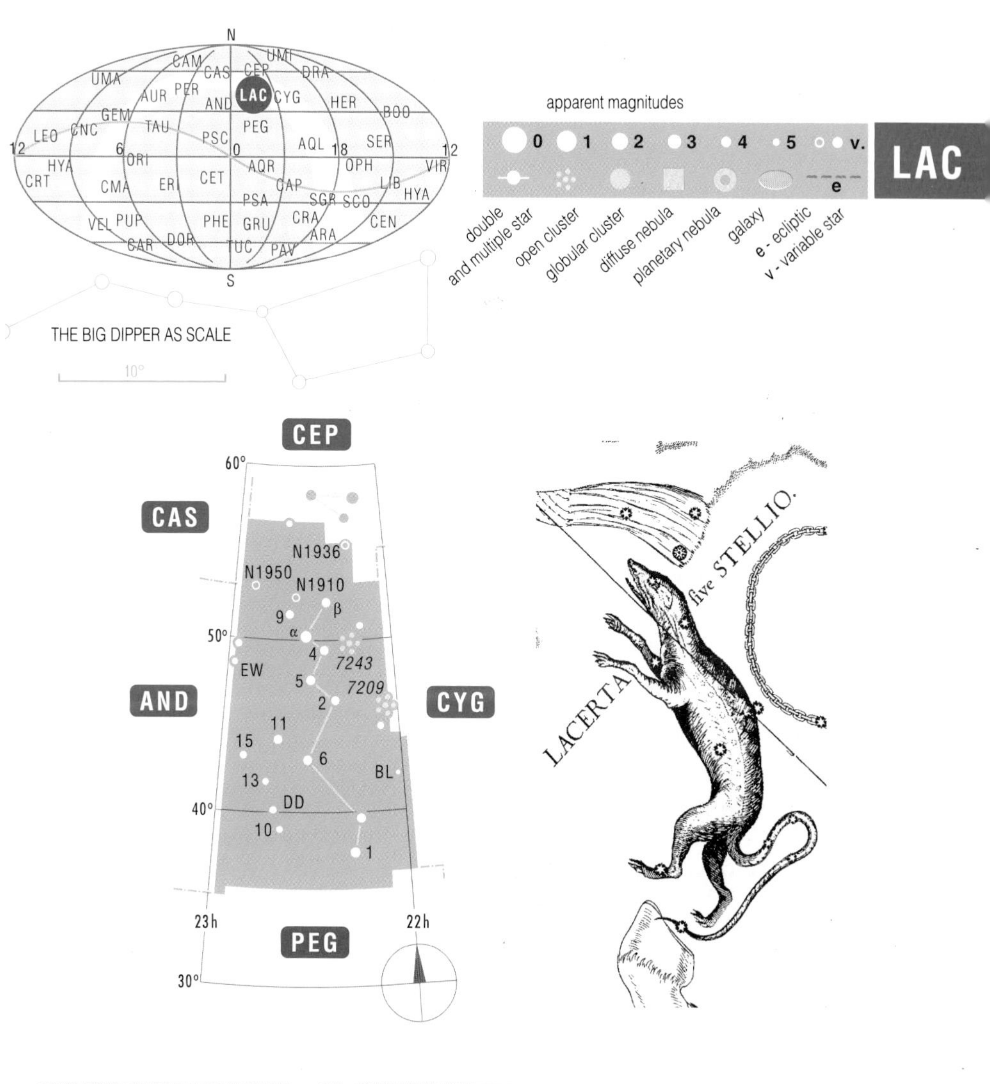

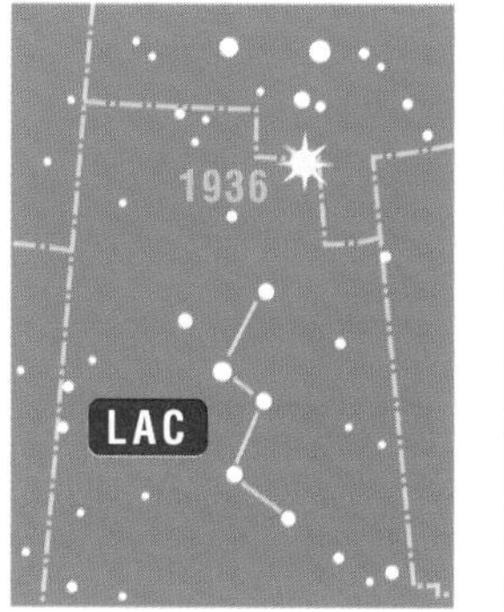

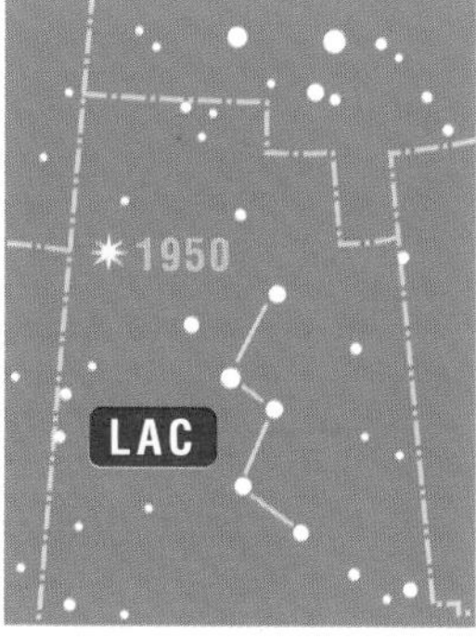

In the 20th century, three novae flared up in Lacerta. The brightest was Nova 1936 (CP Lac) which attained 2.1 mag in its maximum; previously it had only 16.6 mag. Nova 1910 (DI Lac) attained 4.3 mag. The faintest of them was Nova 1950 (DK Lac), which brightened to 5.0 mag.

LEO

Leonis Leo The Lion

The constellation of the zodiac known already in Mesopotamia more than 5,000 years ago. At that time, the Sun stood in Leo at summer solstice, a symbolic expression of the power of the Sun god. According to Greek mythology, it was the dreadful Nemean Lion that was killed by the bold Hercules. The shape of the constellation – one of the few – really corresponds with its name. Its asterism, known as the Sickle, makes up the lion's head and breast.

Alpha Leonis – Regulus, 1.4 mag, is one of the stars of the "spring triangle." It is situated only 0.5° from the ecliptic, the Moon and bright planets passing in its proximity. Regulus is about twice as big as the Sun with the luminosity 120 times higher than that of the Sun. It is a blue-white star of spectral class B7 with surface temperature of 14,000 K, situated at a distance of 78 light-years. **Gamma Leonis – Algieba** ranks among the most beautiful binary stars. Its components of 2.4 and 3.6 mag are separated by 4.5". Both of them are golden yellow and orbit each other about every 600 years. The distance of the double star is 126 light-years. **R Leonis**, a red giant, ranks among long-period variable stars of Mira Ceti type. It changes its brightness within the limits of 4.4 and 11.3 mag in the course of 310 days. The Lion contains numerous bright galaxies. For minor telescopes, two pairs are particularly interesting. **M 65** and **M 66 – NGC 3623** and **NGC 3627**, of about 10.0 mag, are situated at a distance of some 30 to 40 million light-years. The other couple – **M 95** and **M 96 – NGC 3351** and **NGC 3368** – belong to the same group.

LEO MINOR

Leonis Minoris LMi The Lesser Lion

The constellation was introduced by Johannes Hevelius in his atlas of 1690. He placed it between the Lion (Leo) and the Great Bear (Ursa Major) to fill the empty cage of the celestial zoo with some animal of similar character. The constellation has only seven stars brighter than 5.0 mag. Only one star bears Bayer's letter – Beta. All others are denominated with Flamsteed's numbers. There is nothing too interesting about this constellation.

Beta Leonis Minoris, 4.2 mag, is a double star with components of 4.4 and 6.1 mag at mutual apparent distance of only a few tenths of a second. The distance of the system is 146 light-years.

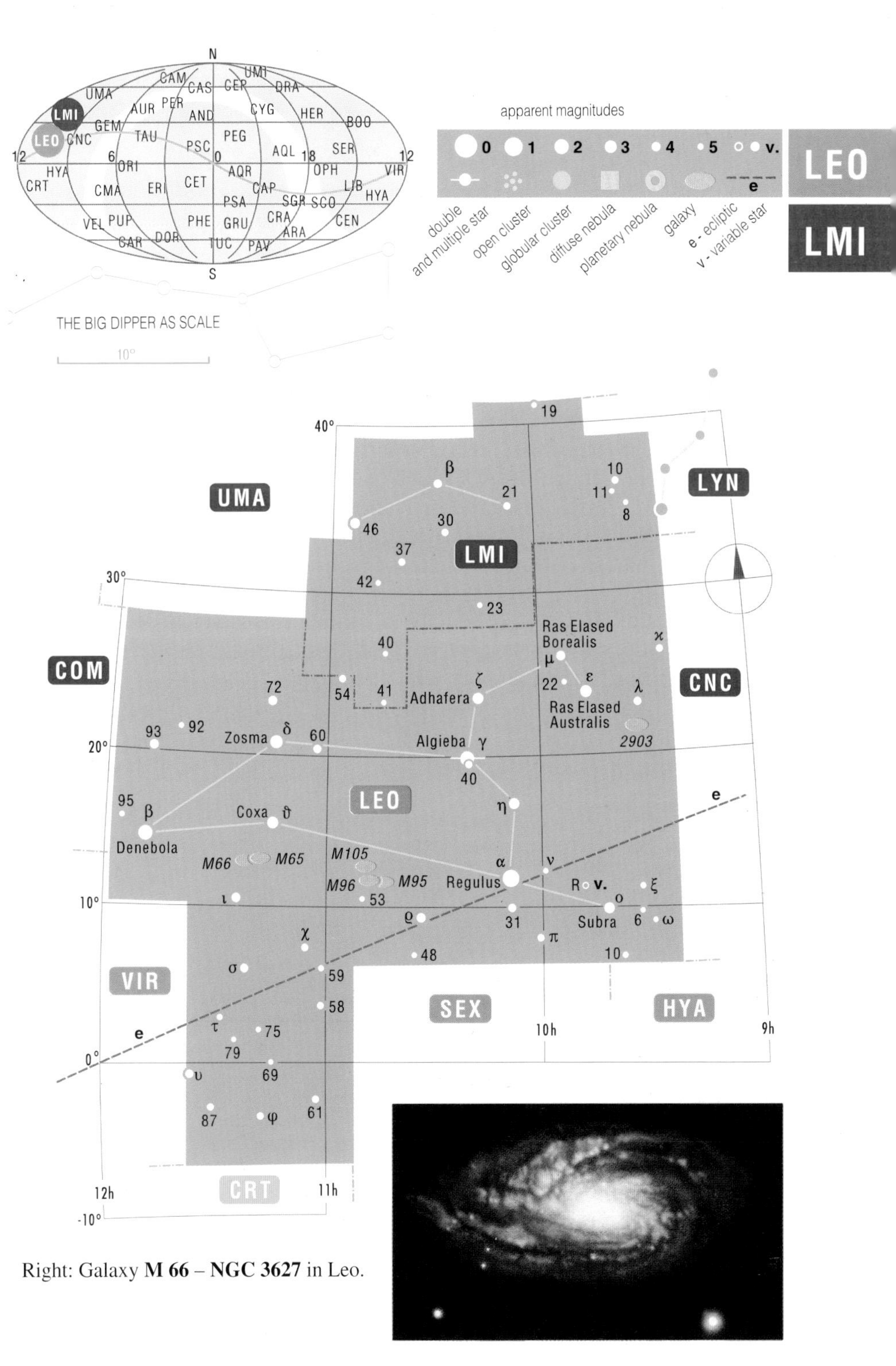

Right: Galaxy **M 66 – NGC 3627** in Leo.

LEPUS

Leporis Lep The Hare

This constellation was known to the ancient Greeks and Romans. The hare, an animal frequently hunted, is placed in the sky at the feet of Orion, the big hunter, within reach of the hunter's dog. The relatively distinct constellation, consequently, can be found easily in the southern neighborhood of Orion, not far west of Sirius.

Alpha Leporis – Arneb, 2.6 mag, is a supergiant 6,000 times as bright as the Sun, which we can see from a distance of more than 1,000 light-years. Alpha has a companion of 11.0 mag at an apparent distance of 36".

R Leporis – Hind's Crimson Star, an irregular variable star. The British astronomer John Russell Hind described it in 1845 as an "intensively crimson star recalling a drop of blood on the background of the sky." The variations of its brightness are characterized by its light curve (below). R Leporis is a giant star of low surface temperature of about 2,700 K. Its atmosphere contains carbon molecules, which absorb short-wave (blue) radiation. Moreover, the star is surrounded by a cloud of dust. All this causes the red color of the star – particularly conspicuous during the period of its brightness. The star can be observed with binoculars. However, the color can be discerned distinctly only with a telescope.

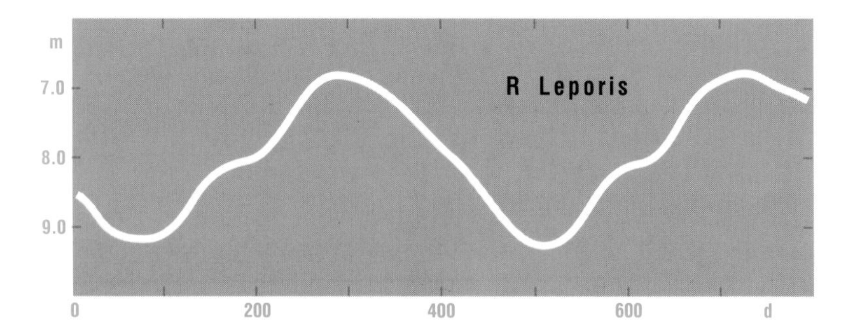

Light curve of the irregularly variable star **R Leporis**. In recent years the star has been changing its brightness from 7.0 to 10.0 mag in two overlapping periods: the shorter lasts about 14 months, the longer several decades (about 50 years).

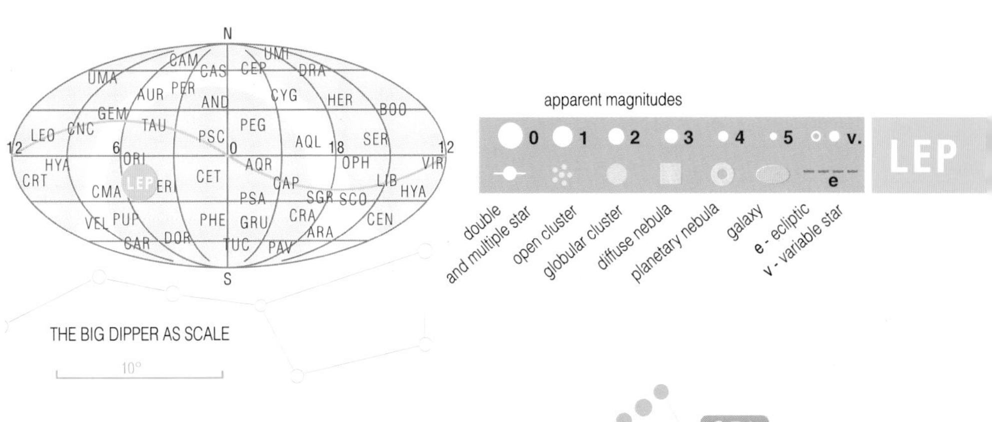

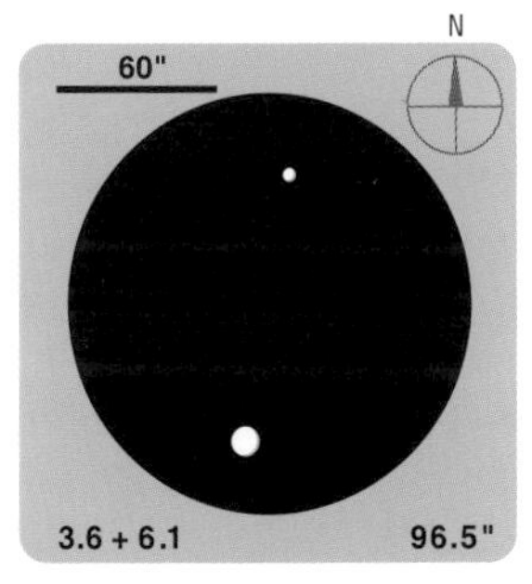

Above: Double star **Gamma Leporis** can be observed with binoculars. It consists of two dwarfs at a distance of merely 29 light-years.

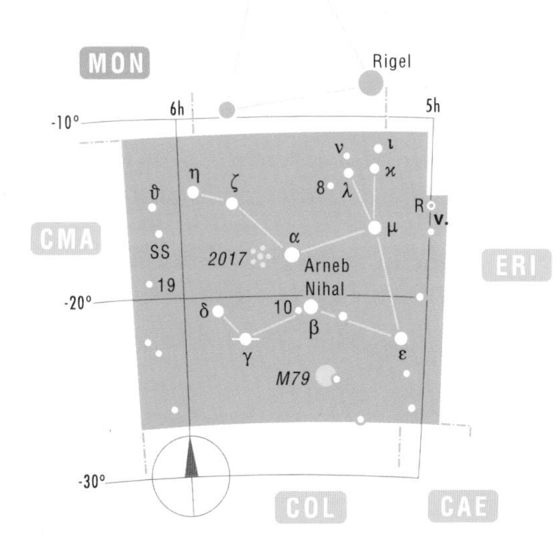

Globular star cluster **M 79 – NGC 1904**, 7.8 mag, apparent diameter of 8′, actual diameter about 260 light-years. Distance from the Sun 42,500 light-years, distance from the center of the Galaxy 63,500 light-years.

LIBRA

Librae Lib The Scales, The Balance

The constellation of the zodiac found easily in the middle of the line connecting the bright stars Spica in Virgo (the Virgin) and Antares in Scorpio (the Scorpion). The names of the bright stars in Libra (the Scales or the Balance) testify to the fact that initially they belonged to the Scorpion: Zubenelgenubi means "the southern claw," and Zubenelschemali "the northern claw." The Sun entered the constellation of Libra, at present situated in the constellation of Virgo (see precession, p. 58) on the day of the autumn equinox – the name represents the equilibrium between day and night and between the seasons. The constellation has been known as the Scales – symbol of justice – since the time of ancient Rome. The pictorial presentation of Libra according to Hevelius' atlas of 1690 is shown on the opposite page.

Iota Librae is a multiple star. Component A, 4.5 mag, is situated at an apparent distance of 58" from component B, 9.4 mag. Component A is a close binary with a companion only 0.1" distant, component B is also a double star with a companion of 11.0 mag at a apparent distance of 1.9". The eclipsing binary **Delta Librae** changes its brightness from 4.9 to 5.9 mag within a period of 2.33 days. Its distance is about 300 light-years.

Globular star cluster **NGC 5897** has an overall brightness of 8.6 mag and an apparent diameter of 8.7′. Its distance from the Sun is 38,500 light-years, and from the center of the galaxy 22,000 light-years.

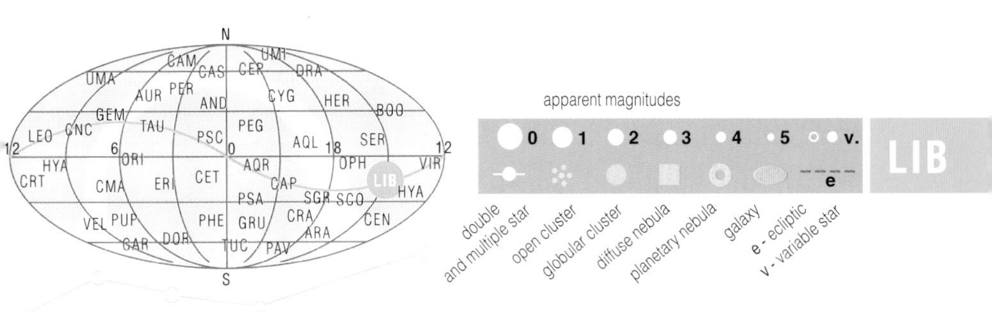

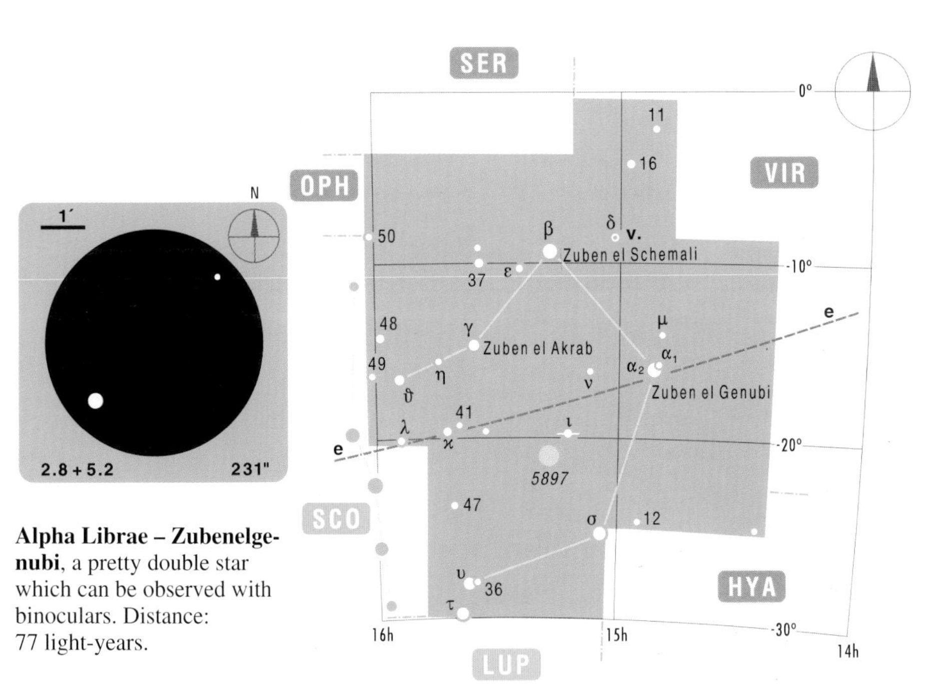

Alpha Librae – Zubenelgenubi, a pretty double star which can be observed with binoculars. Distance: 77 light-years.

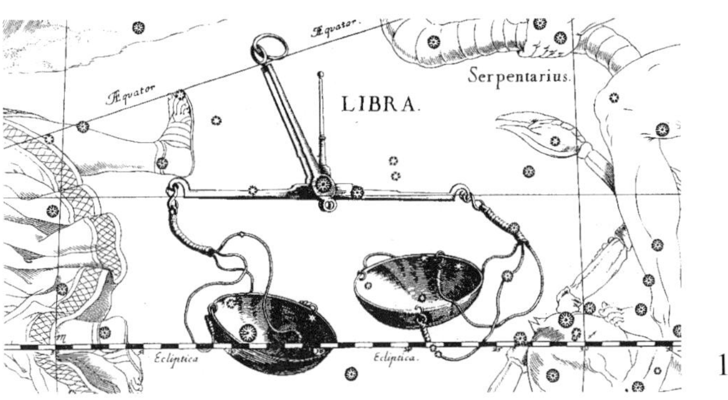

LUPUS

Lupi Lup The Wolf

The Greeks and the Romans saw this group of stars as some sort of wild beast. Later on it was included in pictorial maps as a wolf killed by the Centaur. That is also the way J. Hevelius included the Wolf in his atlas of 1690 – see the picture below. Let us note that Hevelius drew stellar maps – as it was the custom in his period – from the opposite side, as if from "God's view" from the other side of the celestial sphere. For this reason, the Centaur is attacking the Wolf from the left, although we see them in the sky the other way round. The distinct Centaur will also help us to find the Wolf.

The brightest star of the Wolf, **Alpha Lupi**, 2.3 mag, is a double star with a faint companion of 13.4 mag. Apart from the two doubles shown on opposite page let us mention also another double star, **Xi Lupi**, the components of which of 5.3 and 5.8 mag are separated at 10.4".

The open star cluster **NGC 5822** has a total brightness of 6.5 mag and an apparent diameter of 39". It can be observed with binoculars. The globular cluster **NGC 5986**, 7.5 mag, apparent diameter 6', is situated more than 34,000 light-years from the Sun.

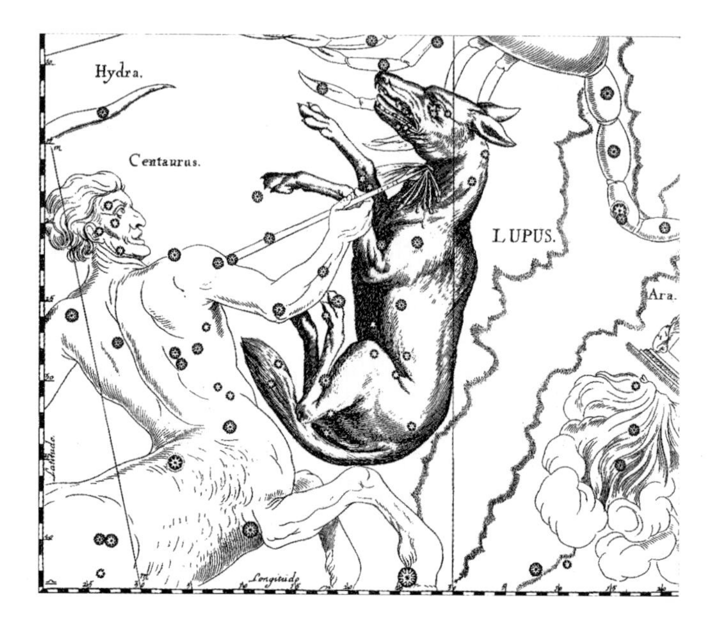

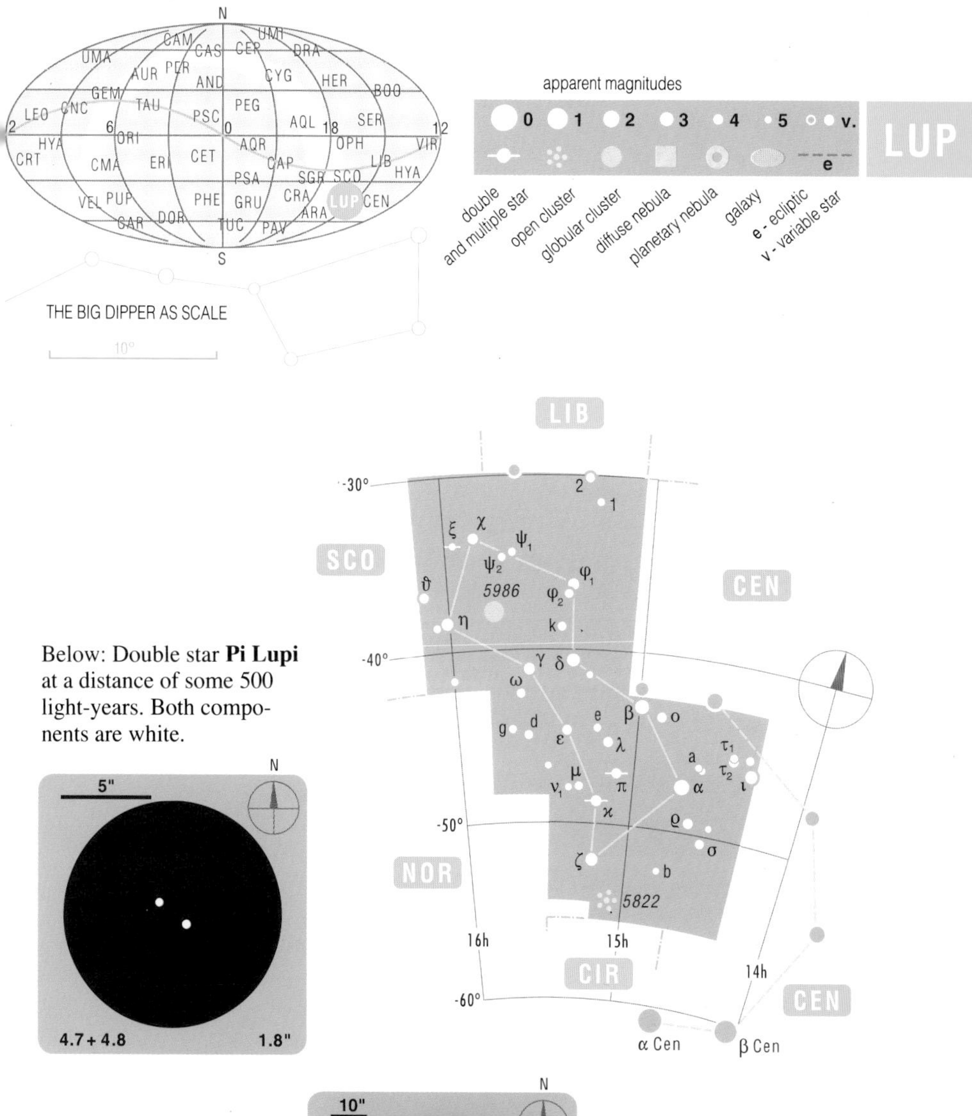

Below: Double star **Pi Lupi** at a distance of some 500 light-years. Both components are white.

10"
N
3.9 + 5.8
26.6"

Left: The components of the double star **Kappa Lupi** can be separated easily with a small telescope. The distance of the system is 190 light-years from the Sun.

LYNX

Lyncis Lyn The Lynx

The constellation of the Lynx was introduced by Johannes Hevelius in 1690 to fill the void between the Great Bear (Ursa Major), the Charioteer (Auriga) and the Twins (Gemini). The lynx has exceptional eyesight and Hevelius indicated symbolically that only people with lynx eyes have the chance of seeing this constellation. The astronomers say that the Lynx is "where there isn't anything." There are only two stars brighter than 4.0 mag, and 12 stars brighter than 5.0 mag. On the map opposite you can find the star **41 Lyncis**, curiously beyond the boundaries of the Lynx in Ursa Major, and, on the other hand, the star **10 UMa** is within the boundaries of the Lynx. It is the heritage of the time before the introduction of definite constellation boundaries.

The globular star cluster **NGC 2419** (see picture below) is very faint (10.4 mag) and has an apparent diameter of only 4'. Its actual diameter is 380 light-years and its luminosity is 175,000 suns. It is the most distant globular star cluster belonging to our Galaxy. It is situated at a distance of 210,000 light-years from the center of the Galaxy and 182,000 light-years from the Sun. It is more distant than the nearest galaxy of Large Magellanic Cloud. Hence its nickname: "Intergalactic Tramp."

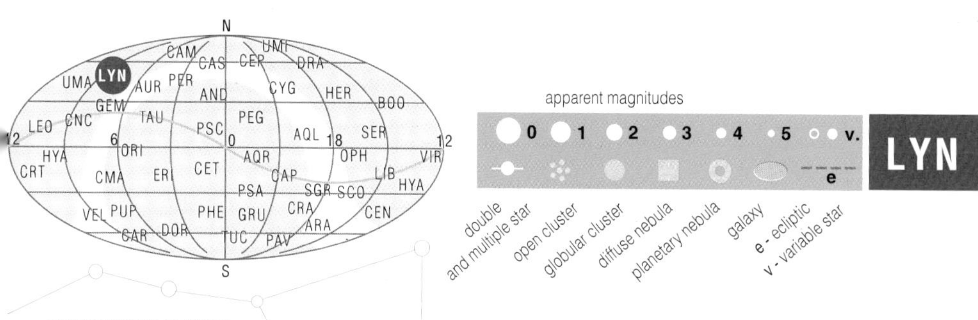

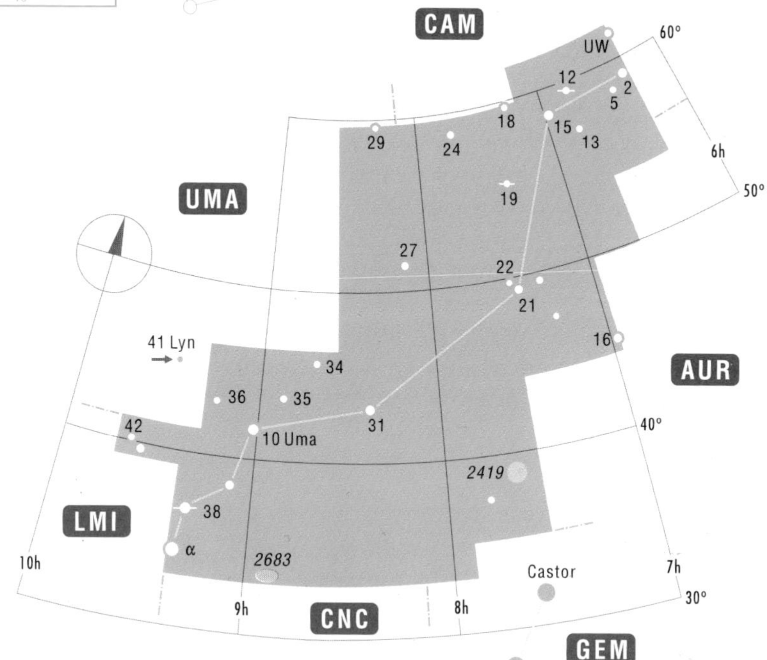

N
5"
5.4 + 6.0
1.7"

Left: In the double star **12 Lyncis**, situated at a distance of 230 light-years, its components revolve about each other in 700 years.

Right: The light from the components of the optical double star **19 Lyncis** travels to us in 360 and 470 years. When observed with the naked eye it appears as a faint little star of 5.4 mag.

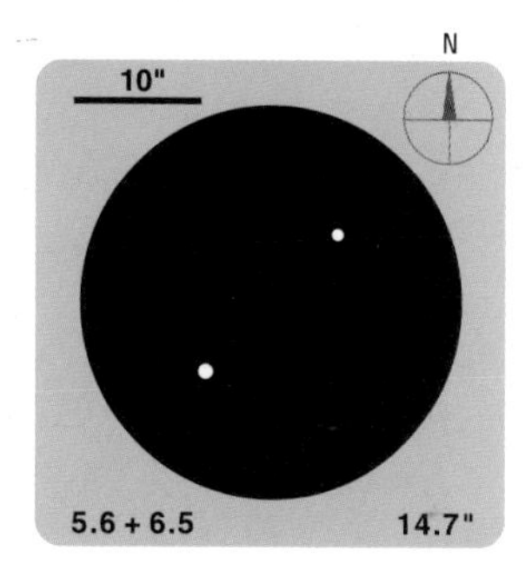

LYRA

Lyrae Lyr The Lyre

Lyra was a stringed instrument which Apollo gave to his son Orpheus. The touching story of Orpheus and his wife Eurydice was put to music by Gluck and Monteverdi. Let us add that after Orpheus' death, Zeus placed the lyre in the sky. The small, but conspicuous constellation can be found best with the assistance of Vega, the brightest star of summer sky.

Alpha Lyrae – Vega, 0.0 mag, the fifth brightest star of the sky, is an ordinary star of spectral class A0 with a higher surface temperature and about three times bigger than the Sun. The light from Vega travels to us in about 25.3 light-years. In 1983, detectors of the Infrared Astronomical Satellite (IRAS) recorded a ring or disc of dust around Vega, which perhaps could be the beginning of a planetary system. **Beta Lyrae – Sheliak** is an outstanding eclipsing binary – the representative of a special type of system in which both stars are so close to each other that gravitational effects and fast rotation have considerably elongated the stars. The changes of brightness of such systems are influenced also by the flow of matter from one component to the other and to the ambient space. In its maximum, Beta is of 3.3 mag, its minimum varying between 3.8 mag and 4.1 mag. In a thirteen-day-cycle, 2 maxima and 2 minima alternate. **Delta 1, Delta 2**, 5.6 and 4.3 mag (spaced 620" apart), is a double star that can be observed with binoculars. It is surrounded with a loose star cluster **Stephenson 1**.

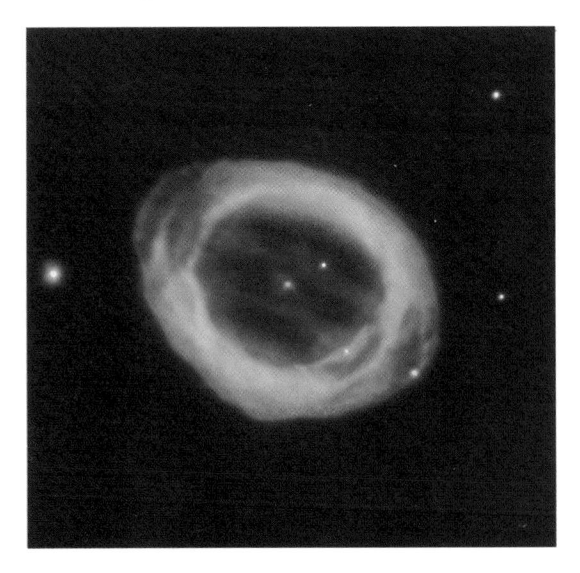

M 57 – NGC 6720 – The Ring Nebula in Lyra, perhaps the best-known planetary nebula. Its total brightness is 9.0 mag, apparent diameter 70", distance about 1,500 to 2,000 light-years. It is not a simple ring, but a complex three-dimensional formation. Its central star is faint (less than 15.0 mag) and cannot be observed with small telescopes.

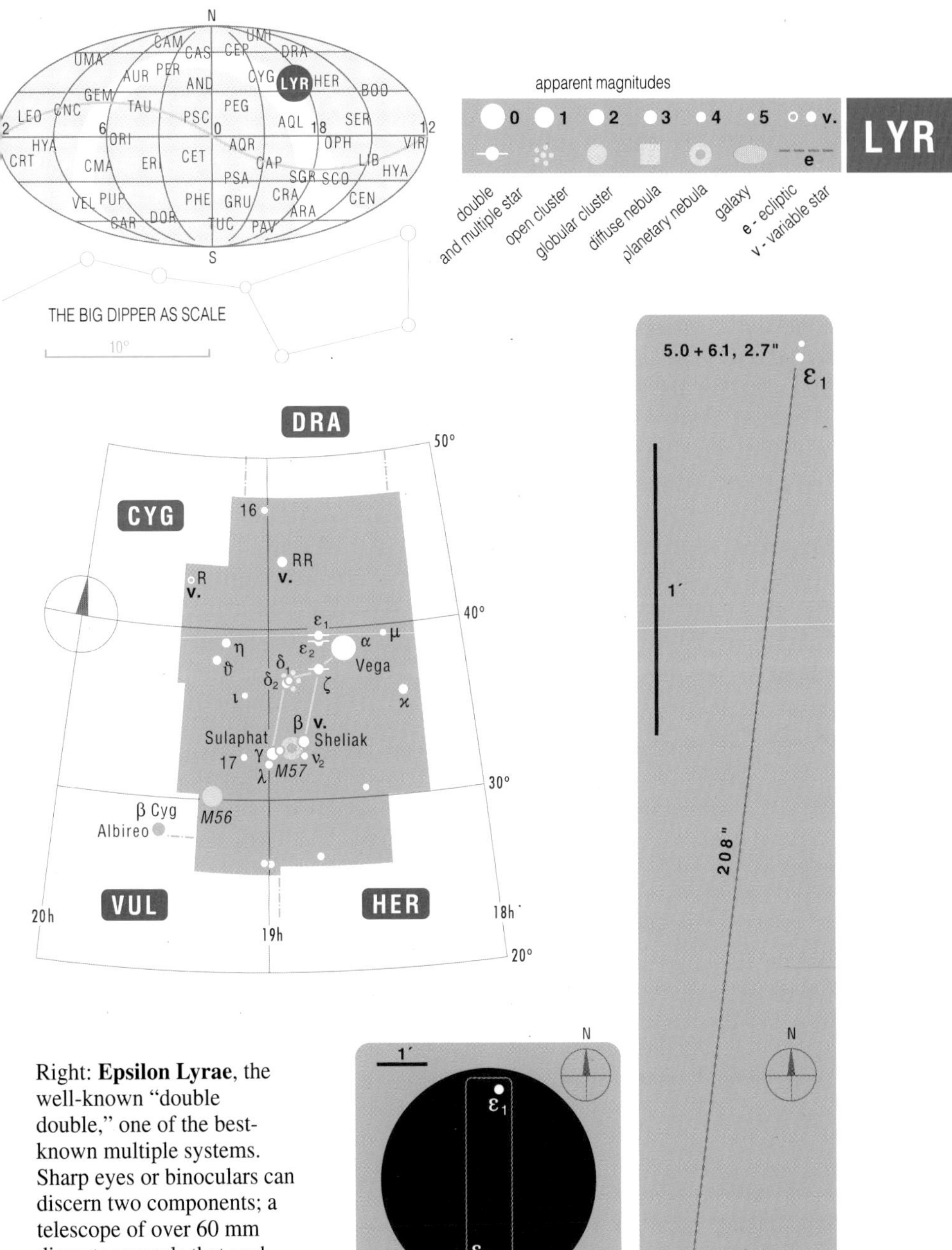

Right: **Epsilon Lyrae**, the well-known "double double," one of the best-known multiple systems. Sharp eyes or binoculars can discern two components; a telescope of over 60 mm diameter reveals that each of them in turn breaks up into two components.

MENSA

Mensae Men The Table Mountain

Once again we come across the name of Nicolas-Louis de Lacaille. This French astronomer filled the gaps among the then known constellations of southern sky. In this area, he immortalized symbolically his temporary observatory below the Table Mountain near Cape Town. Just as the top of the mountain is usually concealed by clouds, the stellar Table Mountain is partly covered by the Large Magellanic Cloud. Mensa is the faintest constellation of the sky in general. There is not one star brighter than 5.0 mag. But for the Magellanic Cloud this area would not attract our attention at all.

The picture of the Table Mountain consists of four stars. The brightest and nearest of them is **Alpha Mensae**, 5.1 mag, at a distance of merely 33 light-years. It is only about 8 light-years farther than the bright Vega. However, Alpha Mensae is much smaller, colder and fainter. It is a dwarf of spectral class G5 and is rather similar to the Sun. **Gamma Mensae,** on the other hand, situated at a distance of 100 light-years, is a giant of spectral class K4 and appears in the sky equally as bright (5.2 mag) as the much nearer Alpha. The stars **Eta** and **Beta Mensae** have apparent stellar magnitude of 5.5 and 5.3 mag. The distance of Eta is about 700 light-years, that of Beta over 600 light-years.

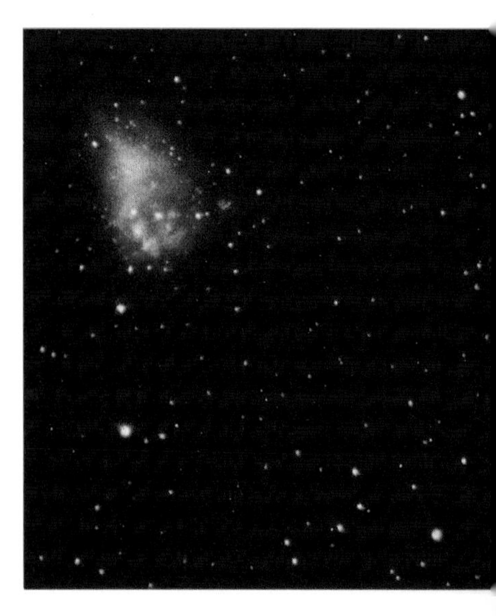

The **Magellanic Clouds** are connected with our Galaxy not only by the gravitational force, but also by a giant gaseous "bridge" of neutral hydrogen. Both Clouds are also mutually connected by a gaseous envelope invisible to human eye. SMC is on the left, LMC on the right.

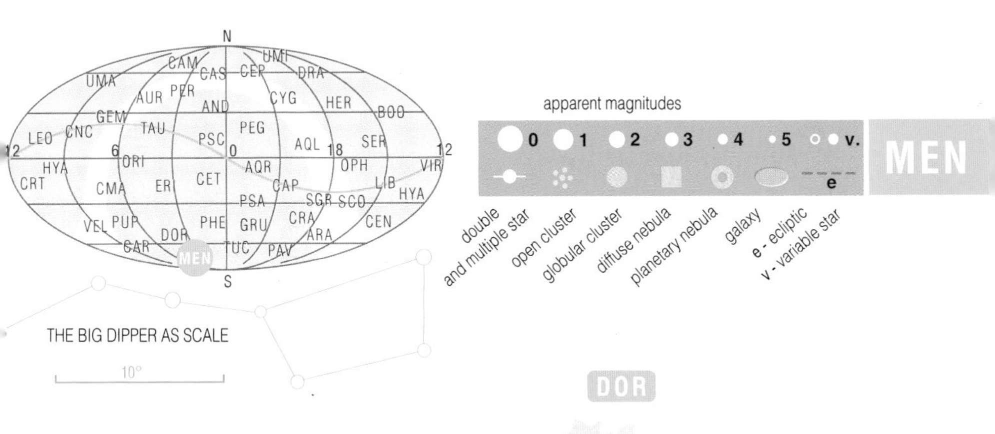

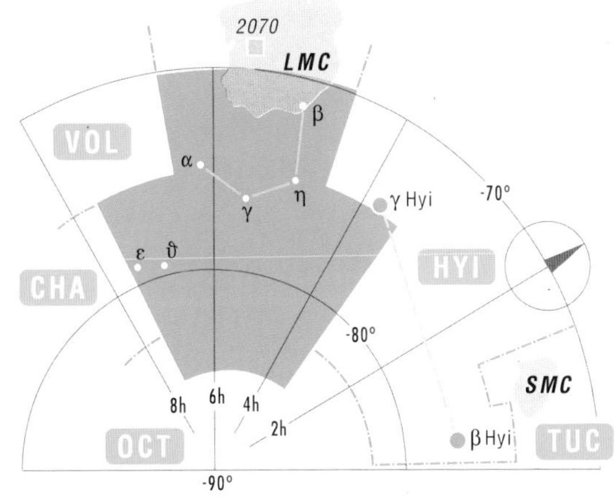

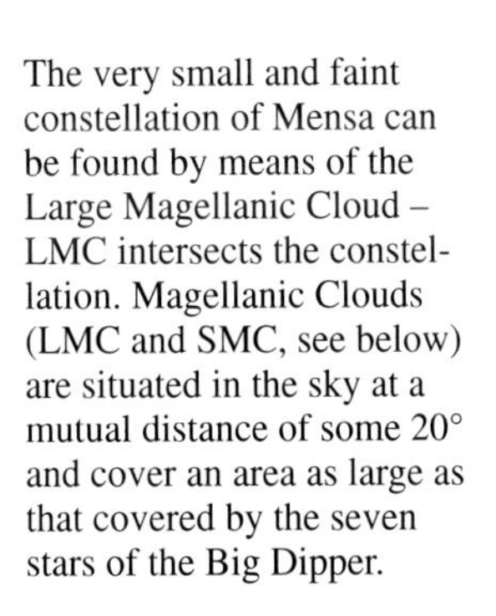

The very small and faint constellation of Mensa can be found by means of the Large Magellanic Cloud – LMC intersects the constellation. Magellanic Clouds (LMC and SMC, see below) are situated in the sky at a mutual distance of some 20° and cover an area as large as that covered by the seven stars of the Big Dipper.

MICROSCOPIUM

Microscopii Mic The Microscope

One of the barely perceptible constellations with which Nicolas-Louis Lacaille filled the voids on the maps of the southern sky in the middle of the 18th century. With this constellation he immortalized the microscope. The modest constellation consisting only of faint starlets is difficult to find south of the Capricorn (the Goat). The initial idea of the microscope is illustrated by a detail of Lacaille's map of the southern sky on the opposite page.

The brightest stars are **Gamma** and **Epsilon Microscopii** of apparent stellar magnitude of 4.7 and a distance from the Sun of 220 and 165 light-years. Gamma is a yellow giant of spectral class G8 with a faint companion of merely 13.7 mag at an apparent distance of 26". Amateur instruments can better observe the not very attractive double star **Alpha Microscopii** illustrated on the opposite page.

Why a Telescope?

To buy or not to buy – that is the question. Do you really need a telescope? What for? For an occasional look at the sky for pleasure, or for some systematic observations of specific character? For a beginner, the market offers innumerable possibilities. Therefore, it is advisable to consult somebody who has some experience with astronomical instruments. In addition to recommendations in the literature, handbooks, instructions for telescope making, periodicals, etc., good advice can be obtained from any planetarium or observatory.

The most important factor is ***not the magnification*** *provided by the telescope, but* ***the quality*** *of its optical equipment and its mounting. Without solid, stable mounting, even the best optics are no good. Beware of highly polished instruments on tottering tripods, offered in department stores!*

If you want a "family telescope," you cannot do better than buy a small refractor (a telescope with an achromatic lens) with an objective 5–7 cm in diameter and a focal length of 50–80 cm. Such instrument permits 50- to 100-fold magnification. according to the eyepiece used. The magnification provided by a telescope can be calculated by the division of the focal length of its objective by the focal length of the eyepiece.

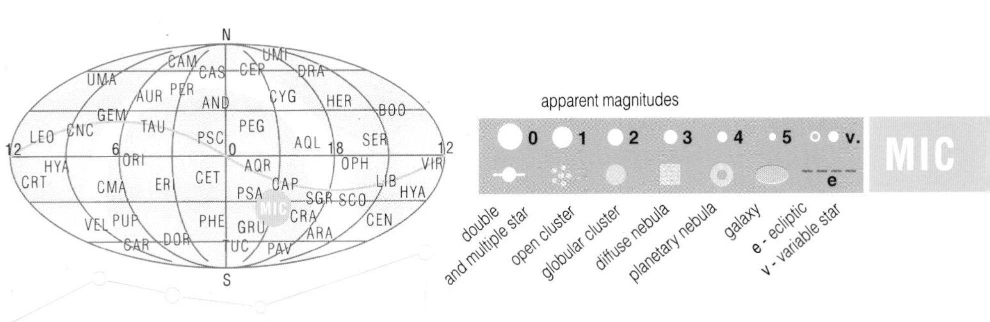

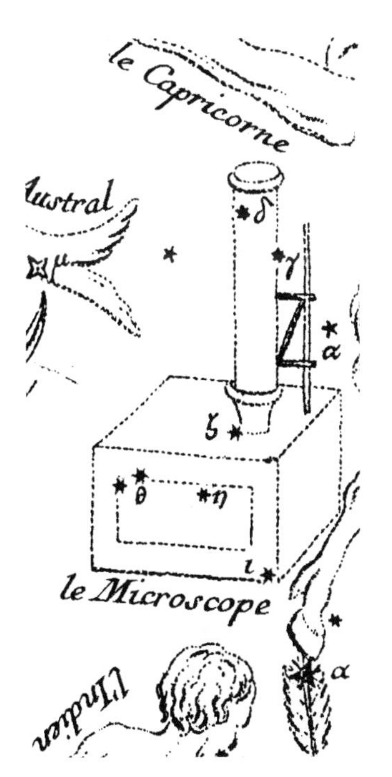

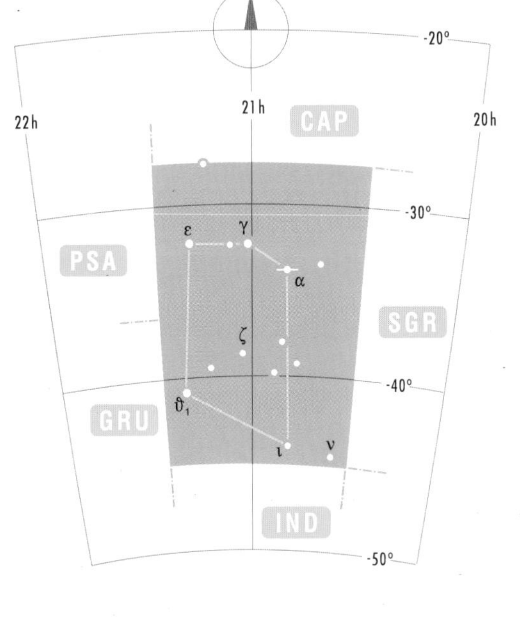

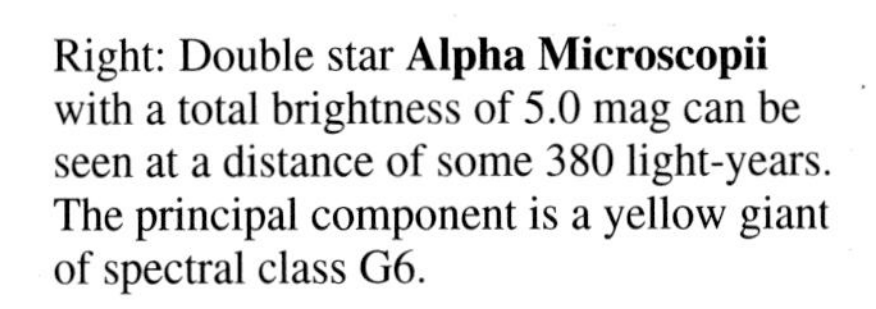

Right: Double star **Alpha Microscopii** with a total brightness of 5.0 mag can be seen at a distance of some 380 light-years. The principal component is a yellow giant of spectral class G6.

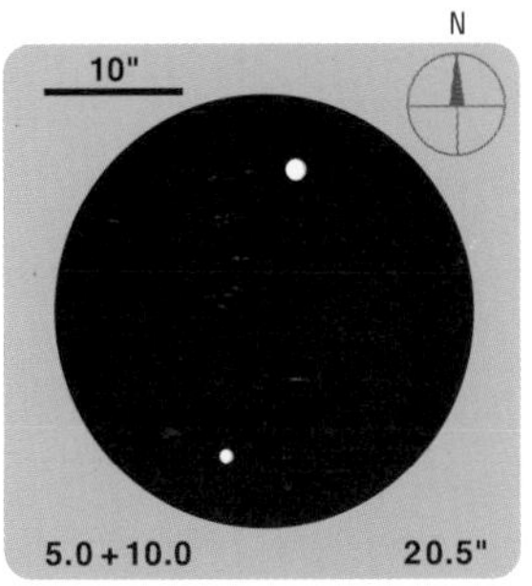

MONOCEROS

Monocerotis Mon The Unicorn

The mythological unicorn used to be depicted as a horse with a long, straight horn projecting from its forehead. This picture was probably inspired by the narwhale. The Unicorn was shown for the first time by Jacob Bartsch (son-in-law of the famous astronomer Johann Kepler) on his map of 1624. It can be found also on Hevelius' stellar map of 1690. The constellation consisting of the stars of the fourth magnitude and fainter is not very prominent in sky, but it can be found easily within the "winter triangle" defined by the bright Procyon, Betelgeuse and Sirius.

The brightest star is **Alpha Monocerotis**, 3.9 mag, an orange giant at a distance of 144 light-years. **Epsilon Monocerotis** is a double star with the components of 4.5 and 6.5 mag at a mutual apparent distance of 13". A star of extraordinary properties is **S Monocerotis**, 4.7 mag, a bluish-white giant of spectral class O7, surface temperature 30,000 K and luminosity of 8,500 suns. The star S Mon is the brightest member of the open cluster **NGC 2264** (see picture below).

The star cluster **NGC 2244**, surrounded by the diffuse nebula **NGC 2237–9 – The Rosette Nebula**, ranks among the most beautiful deep-sky objects (see p. 152). The cluster can be observed even with binoculars, but the nebula is visually faint and stands out only in a photograph.

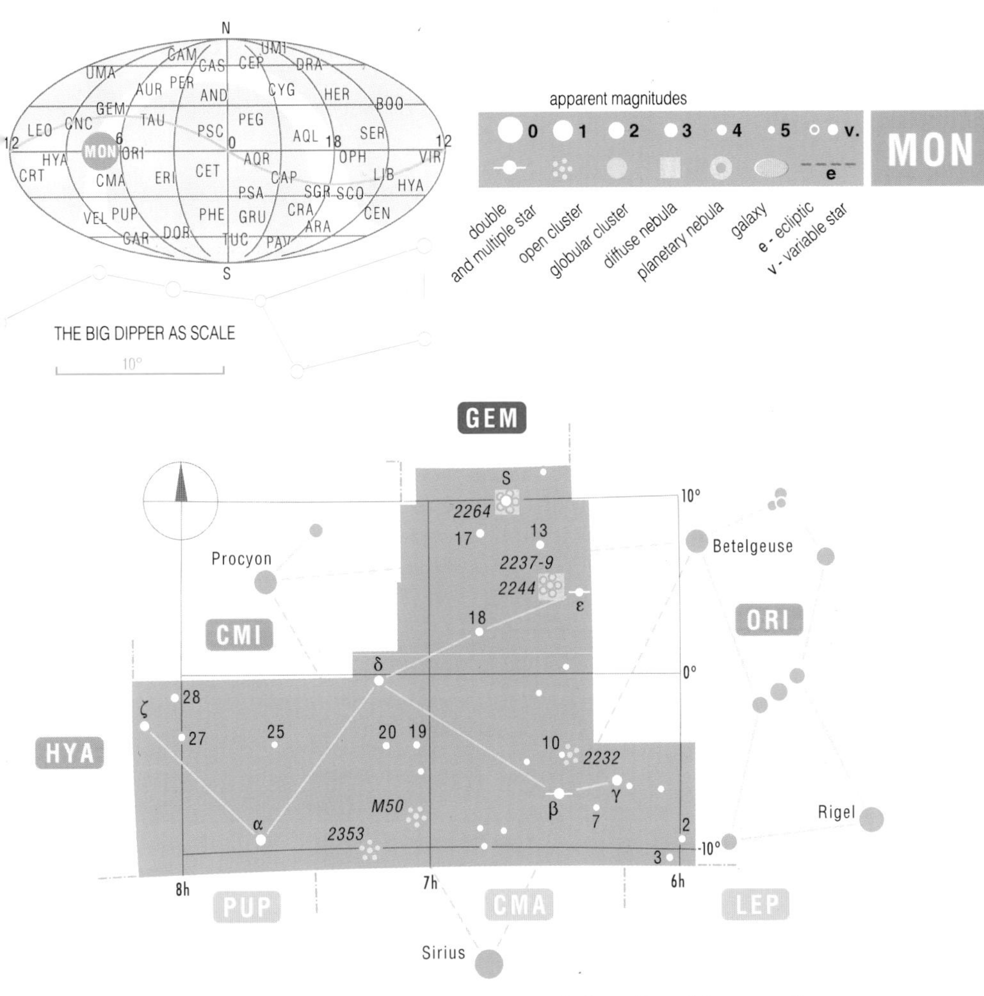

Opposite left: **NGC 2264** – an open star cluster called the Christmas Tree visible with binoculars. Its apparent height is 26′ (almost the diameter of the Moon), actually about 20 light-years. The star cluster is immersed in a nebula which obviously was its cradle. The dark **Cone Nebula** (opposite right) is remarkable. It is a dark cone outlined sharply against the luminous nebula and situated at a distance of some 3,000 light-years.

Below: **Beta Monocerotis** – one of the most beautiful triple stars. A small telescope shows it as a double star.

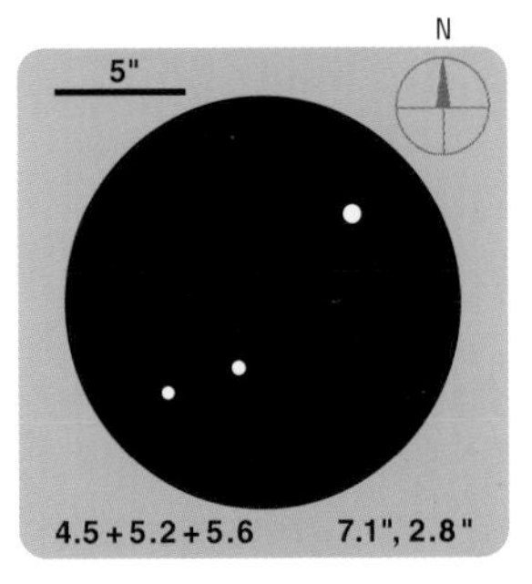

MUSCA

Muscae Mus The Fly

The constellation was introduced by Johann Bayer in his atlas *Uranometria* of 1603, where it was called Apis (the Bee). Later on it was rechristened Musca Australis (the Southern Fly) by Lacaille – not to be confused with the Northern Fly, then shown on the maps near Aries (the Ram). However, the Northern Fly disappeared and its southern counterpart obtained the simple name of Musca (the Fly). Thanks to such an important neighbor as the Southern Cross (Crux), the inconspicuous Fly can be found quite easily.

Alpha Muscae, 2.7 mag, is a double star with a very faint companion of 12.8 mag at an apparent distance of 29.6".

In 1991, a nova (**N 1991** on the map) flared up in the constellation of Musca, which was one of the brightest sources of X-ray radiation in the sky for a short time. We now know that it was a close binary in which an ordinary star of spectral class K revolved about a central object with a mass of over 3 to 8 suns. This object was most probably a so-called black hole, i.e., a high-density collapsed star not emitting any radiation, but manifesting itself by gravitational effects. Its proximity can also produce processes resulting in intensive X-ray radiation.

Diffuse nebula **NGC 2237–9 – The Rosette Nebula**, surrounding the open star cluster **NGC 2244**. The cluster consists of young, hot stars physically connected with the nebula. The diameter of the nebula is about 55 light-years, its distance about 3,000 light-years. (The picture belongs to p. 150.)

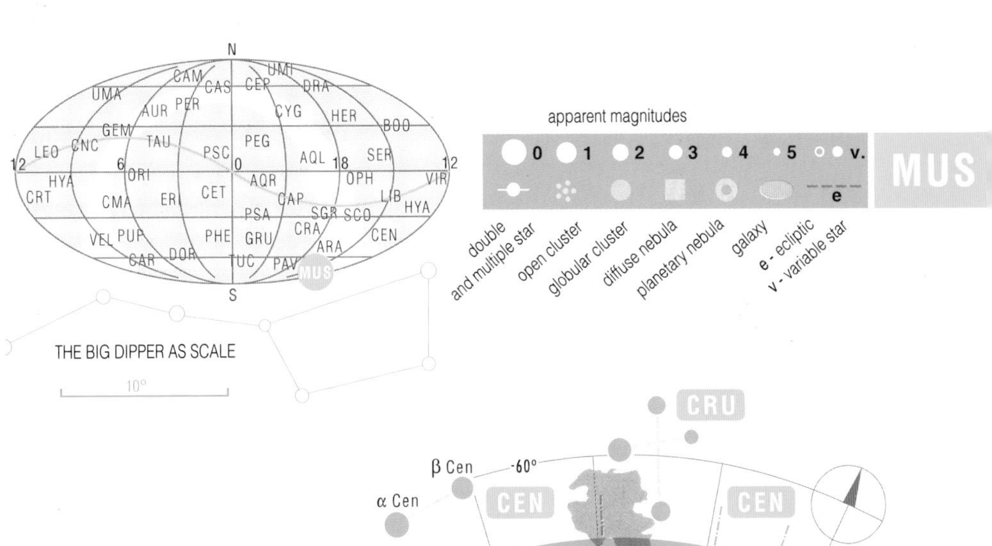

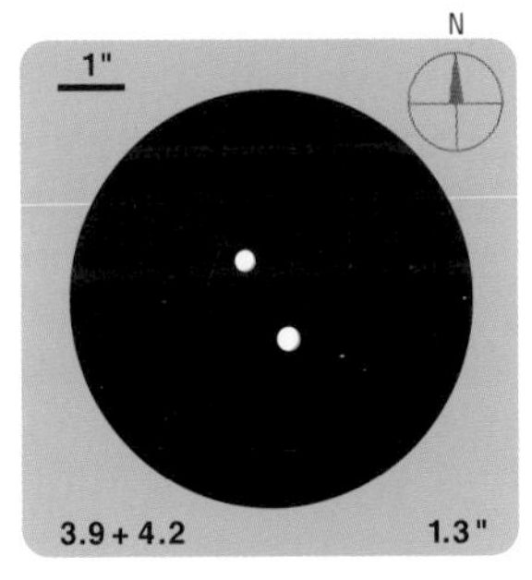

Above: **Beta Muscae** is a binary with a period of revolution of its components of 383 years. Separating the two stars requires a telescope with an objective of more than 100 mm in diameter.

Right: Compare the upper map showing the Fly and the Southern Cross with the photograph of that part of the sky. The dark nebula called the Coal Sack is distinctly outlined against the background of the Milky Way.

NORMA

Normae Nor — The Rule

One of the 14 small constellations introduced by Lacaille to his map of the southern stellar sky in the middle of the 18th century. Since its origin this – one is tempted to say superfluous – constellation has changed its name and boundaries several times. Initially it was to immortalize the craftsman's aids – the rule of the level and carpenter's square. We cannot find Alpha and Beta Normae which were severed from the constellation after the definite boundaries were defined in 1930.

The stars Gamma 1 and Gamma 2 do not form a physical double system. **Gamma 1 Normae**, 5.0 mag, is a white supergiant at a distance of over 1,400 light-years. **Gamma 2 Normae** is "only" 128 light-years away. It is a double star, the main component of which of 4.0 mag is situated at an apparent distance of 45" from its faint secondary component of 10.0 mag. The components of the double star of **Iota 1 Normae** of 4.9 and 8.1 mag are situated at a mutual distance of 10.8".

The nebula **NGC 6164–5** (picture below) has an appearence of a planetary nebula. It has also a central star of 6.8 mag. The object looks like an inverted S, but its actual spatial shape is unknown.

The nebula **NGC 6164–5** is a faint object which can be perceived well on a photograph only. Its apparent diameter is 370". It is situated at the boundary with the adjacent Altar (Ara).

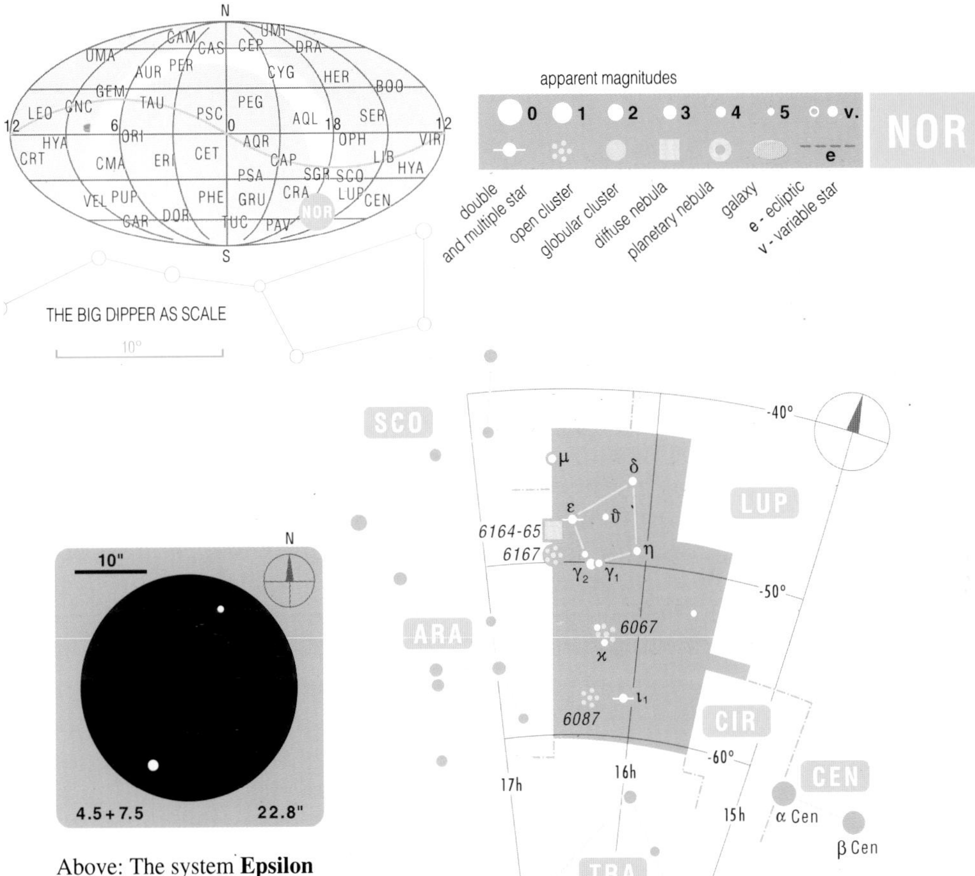

Above: The system **Epsilon Normae** is a quadruple star. Both components visible with the telescope are spectroscopic binaries.

Right: The open star cluster **NGC 6087** can be observed with binoculars. Its total luminosity is 5.4 mag, apparent diameter 12′ and distance about 2,900 light-years. It contains about 40 stars between 8.0 and 10.0 mag.

OCTANS

Octantis Oct The Octant

With this inconspicuous constellation Lacaille memorialized an important navigation instrument invented by John Hadley in 1730. Its initial name was Octans Hadleianus. As the predecessor of the sextant, the octant was also used for the measurements of apparent altitude of celestial bodies above the horizon.

The Octant contains the **south celestial pole**, but has no bright star to indicate the position of this important point like the North Star in northern sky. The role of the hardly visible guard of the pole is played by the faint star of 5.5 mag, **Sigma Octantis**. As shown below, it is only a temporary role thanks to the precession of the Earth's axis, which makes the pole travel along a nearly circular orbit with the center in the pole of the ecliptic (P.E.). Sigma was nearest to the pole in about 1870. At present, it is gradually discarding its role. In the year 2000, its apparent distance from the south pole will be about 1°. The population of the southern hemisphere will enjoy the brightest polar stars between the years 5000 and 11 000, when the pole will pass through the constellations of Carina (the Keel) and Vela (the Sails). In the period from 8000 to 9000, the "false cross" will be a magnificent pole indicator.

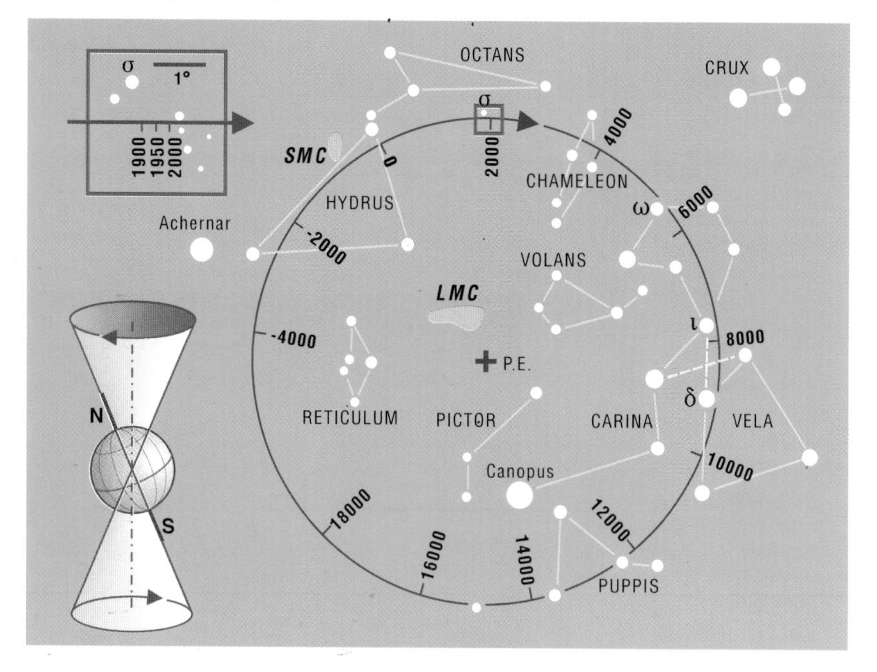

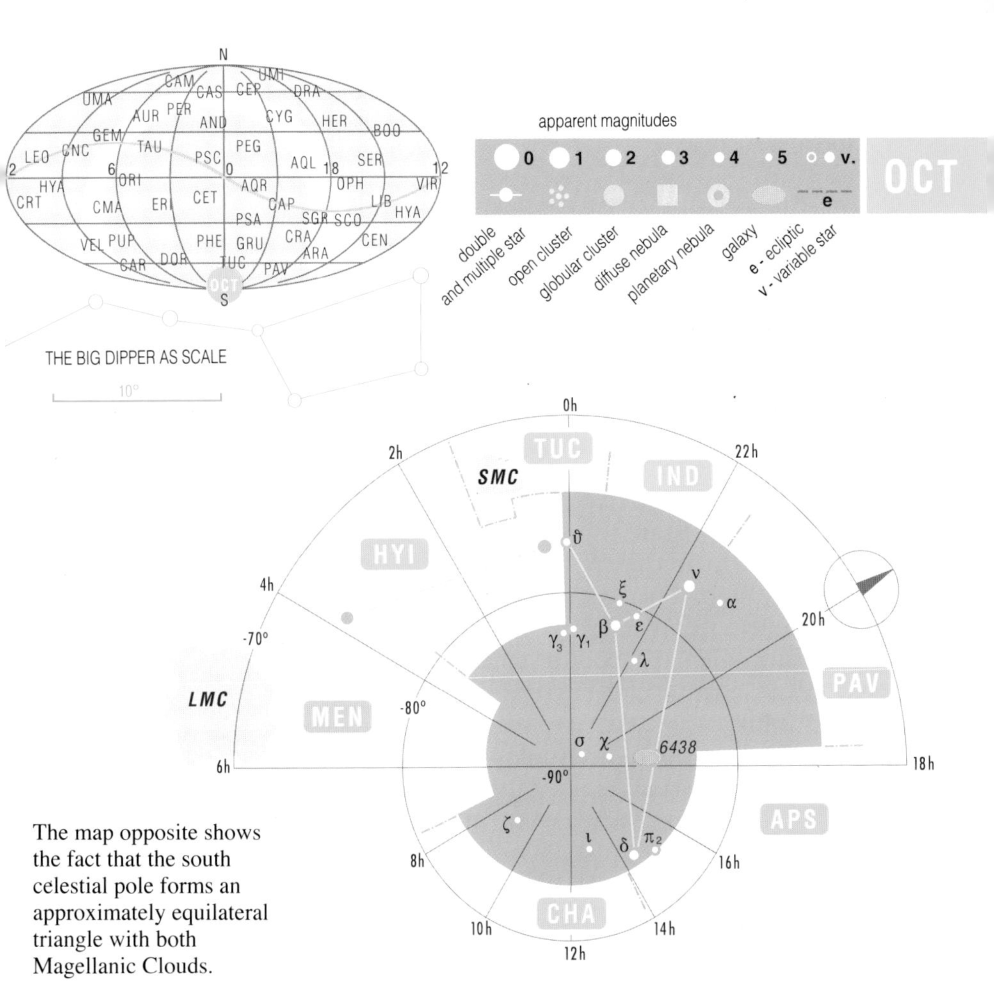

The map opposite shows the fact that the south celestial pole forms an approximately equilateral triangle with both Magellanic Clouds.

The pair of galaxies **NGC 6438–6438A**, situated at a distance of some 110 million light-years and probably linked by gravitational forces. NGC 6438 is of S0 type, while NGC 6438A is an irregular giant galaxy with two broad arms.

OPHIUCHUS

Ophiuchi Oph — The Serpent Bearer

On older maps also called Serpentarius. According to mythology, Asclepius, the god of medicine, had learned the secret of the therapeutic effects of plants from a serpent. He was alleged to be able to bring people back from the dead. Therefore Hades, the god of the Nether World, asked Zeus, his brother, to kill Asclepius by lightning. Zeus immortalized Asclepius and his serpent in the sky. The Serpent Bearer does not belong and has never belonged among the constellations of the zodiac, although the sector of the ecliptic passing through it is three times as long as that of the adjacent Scorpion. In 1604, the so far last supernova of our Galaxy flared up in the Serpent Bearer. It is known as **Kepler's Star**.

Barnard's Star, also known as Velox Barnardi, is a faint red dwarf of 9.5 mag, a luminosity of 1/2,500 sun, and a diameter of about 225,000 km. It is the second nearest star, its distance being only 5.9 light-years. The star holds a record: it has the fastest known proper motion, shifting against the background of more distant stars by one Moon diameter every 175 years. It is approaching us, and in 9,700 years, it will pass us at a distance of 3.8 light-years. At that time, it will be near the Dragon's head in the sky. **Rho Ophiuchi** is a triple star, which can be observed with binoculars. Its components are of 5.0, 7.0 and 8.0 mag. Each of the fainter components is situated at a distance of 2.5′ from the component of 5.0 mag.

Below left: Apparent orbit of the double star **70 Ophiuchi**. The separation of its components varies between 1.7" and 6.7" every 88 years.

Below right: One of the numerous star clusters in Ophiuchus: **M 10 – NGC 6254**, 6.5 mag, apparent diameter 12′, visible with binoculars.

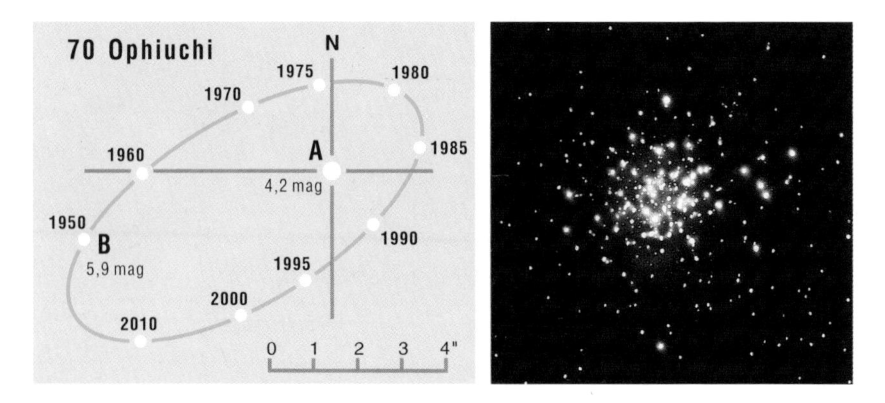

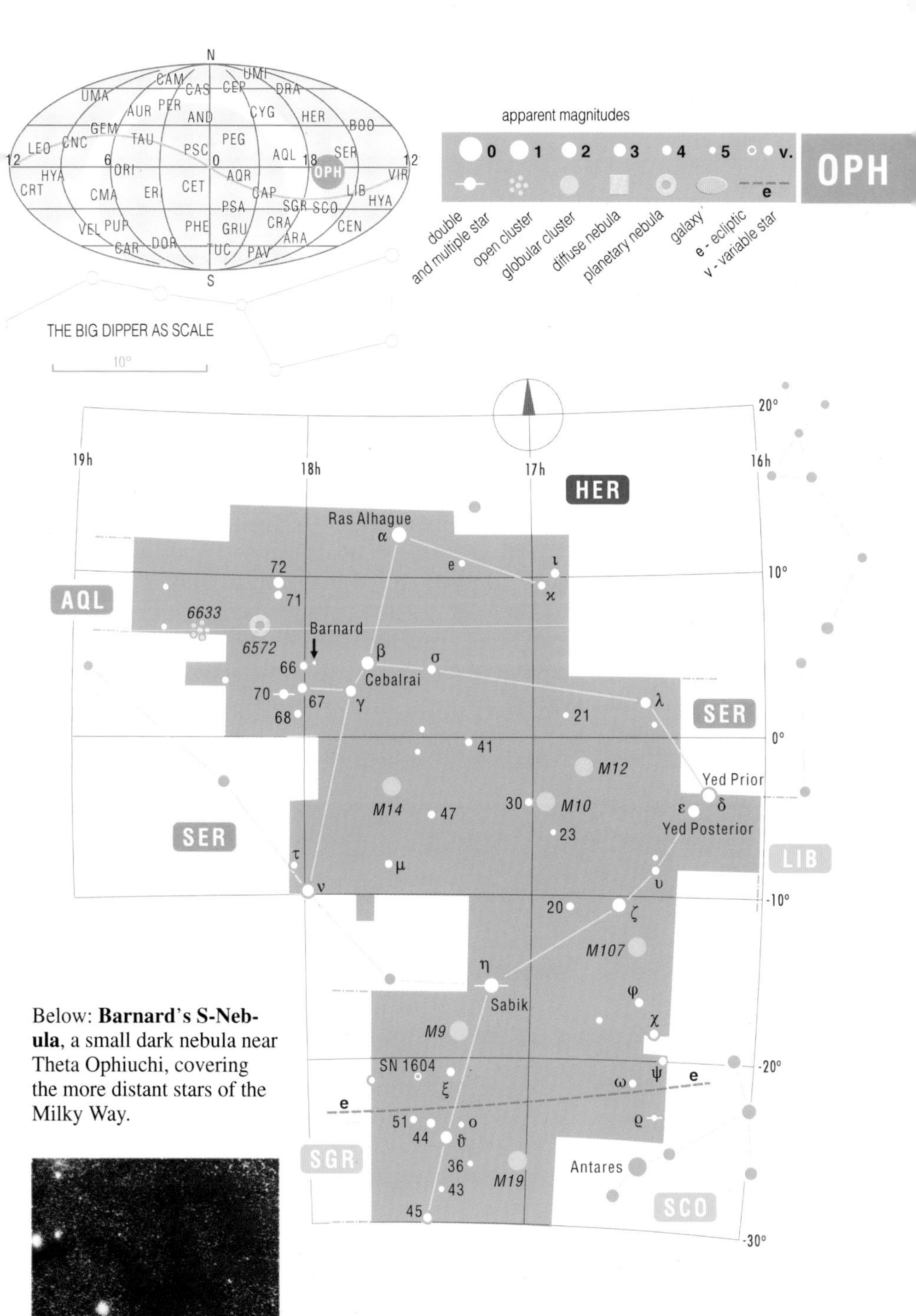

Below: **Barnard's S-Nebula**, a small dark nebula near Theta Ophiuchi, covering the more distant stars of the Milky Way.

ORION

Orionis Ori — Orion, The Hunter

Orion, the great hunter, is undoubtedly the most spectacular constellation of the sky. The stars of Betelgeuse and Bellatrix shine brightly on his shoulders, Rigel and Saiph represent his feet and the unique threesome of Alnitak – Alnilam – Mintaka adorn his belt. The hunter used to be depicted struggling with the Bull (Taurus). According to one myth, Orion was killed by the Scorpion (Scorpius). No wonder they do not appear together above the horizon: when Orion rises, the Scorpion sets, and vice versa.

Alpha Orionis – Betelgeuse – is an irregularly pulsating red supergiant of distinct yellow-orange hue. Its brightness varies from 0.3 to 0.6 mag. It is one of the biggest stars, its diameter changing between 550 and 900 times the diameter of the Sun. **Beta Orionis – Rigel**, 0.1 mag, is one of the most luminous stars: its luminosity equals that of 55,000 suns.

The best-known, most admired and most frequently photographed is the diffuse nebula **M 42 – NGC 1976** (below) – the Orion Nebula. It is visible to the naked eye, binoculars, telescope or in photographs and always exquisite. It is an enormous complex of interstellar gas, dust and molecules of a diameter of some 30 light-years, situated at a distance of 1,600 light-years.

In the brightest nucleus of the nebula **M 42** four stars of 5.4 to 8.0 mag, known as the **Trapezium**, can be observed with even a small telescope. These young and very hot stars are the source of ionizing radiation, making the gas in the nebula optically visible.

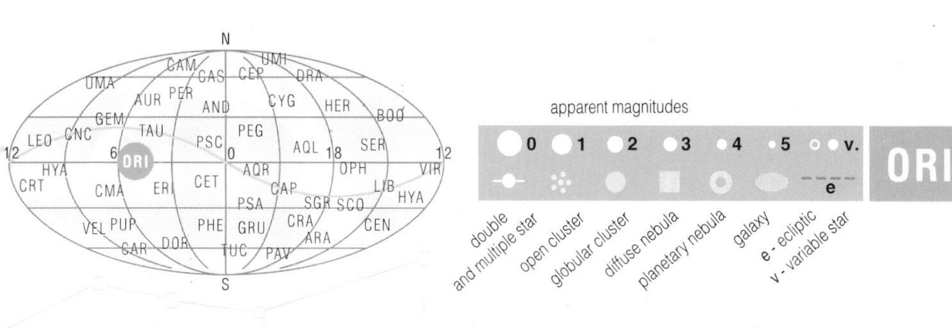

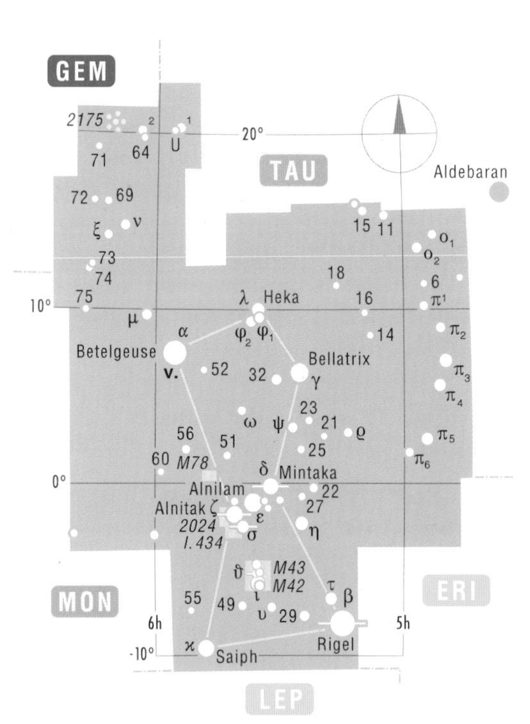

Right above: Complex of nebulae near Zeta Orionis. In the upper left hand corner there is the diffuse nebula **NGC 2024**. Lower on the right, against the background of the faintly shining nebula **IC 434**, the characteristic dark nebula, the **Horsehead Nebula – B 33**, stands out (see the detail, right). The height of this peculiar "chess knight" is about one light-year.

PAVO

Pavonis Pav The Peacock

The Peacock is one of the twelve "modern" constellations in the southern sky introduced by Johann Bayer in his atlas *Uranometria* in 1603 according to the proposals of the Dutch navigator P. D. Keyser. He probably had in mind a certain peacock of Greek mythology, the tail plumes of which the goddess Hera decorated with the hundred eyes of the giant Argus. The poor giant was the victim of a dispute of Zeus and Hera over one of the innumerable amorous adventures of the supreme thunder-wielding god.

Alpha Pavonis – Peacock, 1.9 mag, is the brightest star not only of the constellation, but also in its broad environs. Therefore, it is an excellent orientation point. It is a multiple star with companions of 9.0 and 10.3 mag. The main star is of spectral class B2. Its diameter is 2.5 times the diameter of the Sun, its brightness is more than 1,000 times that of the Sun. Its distance is 183 light-years. **Kappa Pavonis** is one of the brightest cepheids (see pp. 24, 84) changing brightness from 4.0 to 4. 9 mag within 9.1 days.

The pair of galaxies **NGC 6769** and **NGC 6770** (opposite below) is surrounded by a common envelope. The galaxies influence each other by gravitational forces. Their distance is about 190 million light-years.

Globular star cluster **NGC 6752**, total brightness 5.5 mag, apparent diameter 42', distance from the Sun about 14,000 light-years.

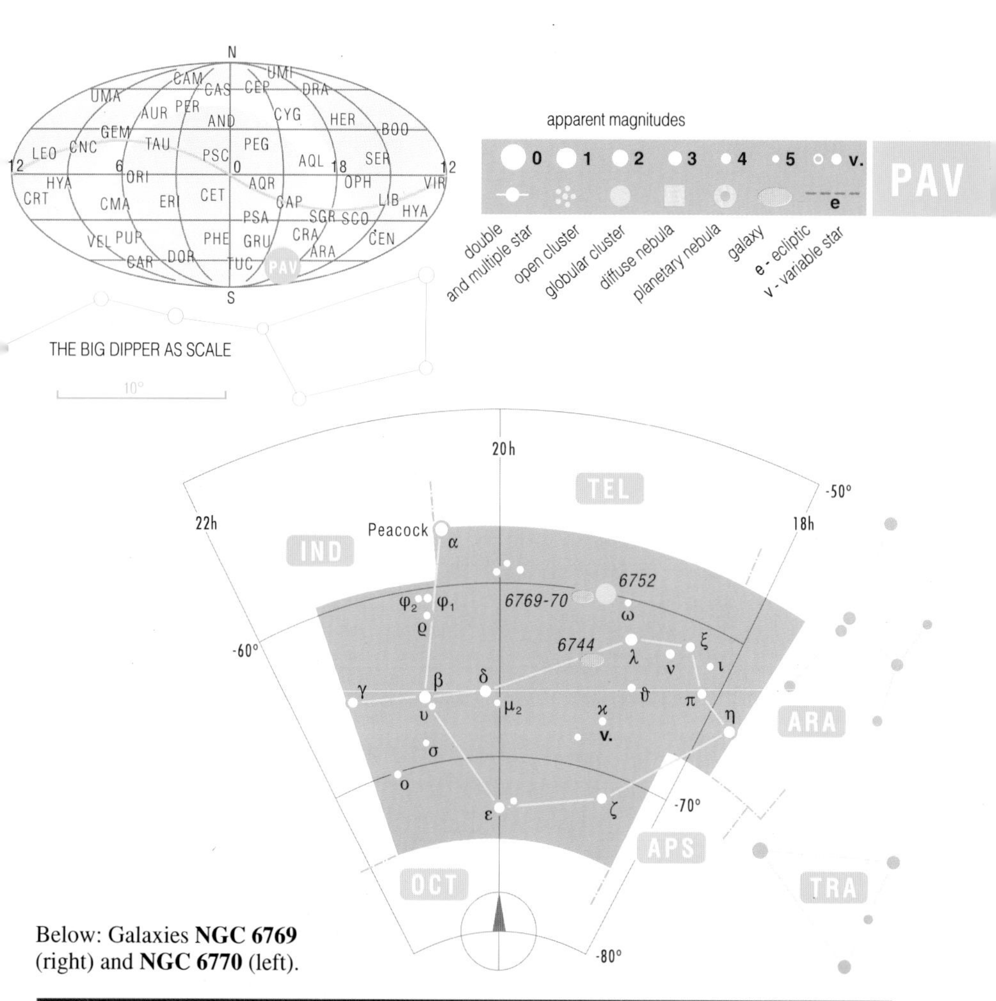

Below: Galaxies **NGC 6769** (right) and **NGC 6770** (left).

PEGASUS

Pegasi Peg Pegasus

According to mythology, Pegasus, the winged horse, flew out of the body of the gorgon Medusa, who was killed by Perseus. The horse is said to have been tamed by the hero Bellerophon, who killed the monster Chimaera with his aid subsequently. According to another legend, the first rider of the winged horse was Perseus who attacked the sea monster Cetus from the air, thus saving the life of Andromeda. In the sky, the figure of Pegasus is oriented upside down. The horse's back consists of Alpha, Beta and Gamma Pegasi, his head is decorated by Zeta, Theta and Epsilon Pegasi. Alpha, Beta and Gamma Pegasi, together with Alpha Andromedae, form the well known Great Square of Pegasus, a very useful means of orientation in the sky.

Beta Pegasi – Scheat changes brightness from 2.4 to 2.6 mag approximately within 30 days. It is a red supergiant of spectral class M2. Its diameter is estimated at 90 to 110 diameters of the Sun, its distance is 199 light-years. **Epsilon Pegasi – Enif** (Snout), 2.4 mag, has two companions of 8.4 and 11.3 mag. The main star is an orange supergiant shining at a distance of some 670 light-years.

The rich globular star cluster **M 15 – NGC 7078** of total 6.0 mag is well visible with binoculars. **NGC 7479** (below) is an example of a barred galaxy of SBb type. **Stephan's Quintet** is a group of galaxies, some of which are mutually connected by "bridges" of intergalactic matter.

Galaxy **NGC 7479** (left) and group of galaxies known as **Stephan's Quintet** (right).

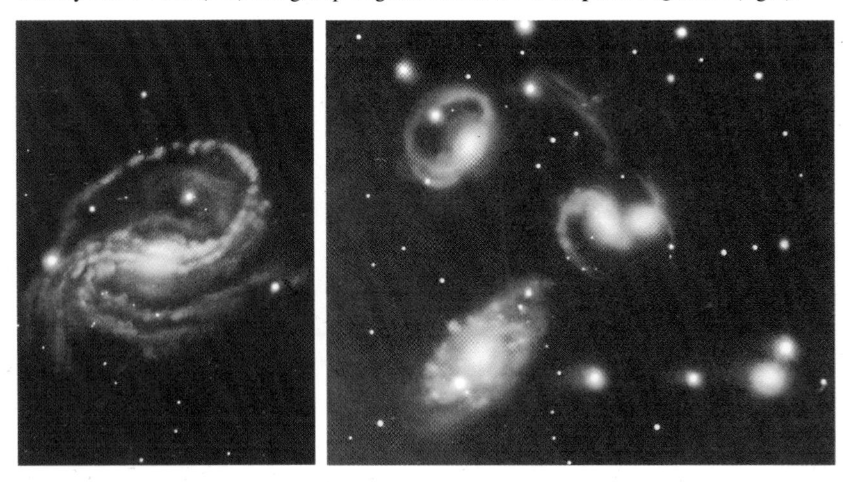

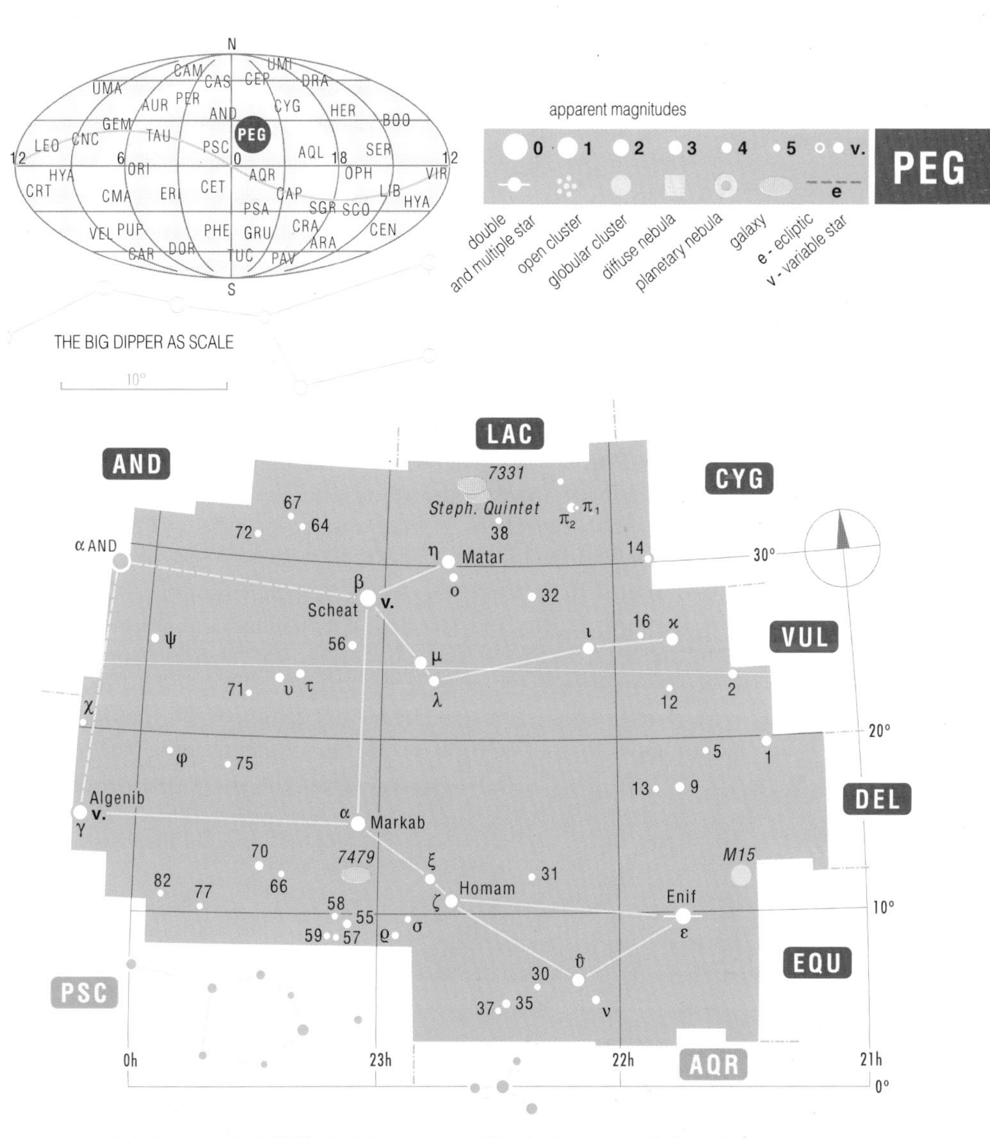

Below: Spiral galaxy **NGC 7331**, 9.0 mag, type Sb, similar to our Galaxy. It is situated some 50 million light-years away, about seventeen times as far as M 31 in Andromeda.

PERSEUS

Persei Per — Perseus

The hero of Greek myths, the son of Zeus and Danae, Perseus belongs to the "royal family" in the sky, together with Cassiopeia, Cepheus and Andromeda. Armed and protected by gods, Perseus beheaded the formidable gorgon, Medusa, from whose blood the winged horse Pegasus originated. When flying back (we do not know, whether on the back of Pegasus or by means of winged sandals) he saw Andromeda chained to the rock and saved her from a certain death in the claws of Cetus, the sea monster.

Alpha Persei – Algenib, **Mirfak**, 1.8 mag, is a yellow supergiant in the middle of a loose open star cluster known as the moving cluster of Alpha Persei – its members move in the same direction. The cluster numbers over 50 stars within a circle of 3° in diameter and affords a nice view to binoculars. The age of the cluster is about 50 million years, its distance some 600 light-years. **Beta Persei – Algol** (Demon Star) is the best-known eclipsing binary (see p. 24), situated at a distance of 93 light-years.

Double star cluster **Chi and h Persei – NGC 884** and **869** (opposite below) can be seen with the naked eye, but displays its splendor only if observed with a telescope. Each star cluster has an apparent diameter of 0.5°, actual diameter of about 60 light-years, and contains some 300 stars. The distance of these star clusters is over 7,000 light-years.

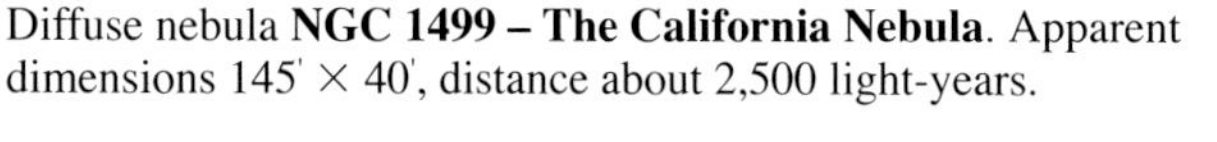

Diffuse nebula **NGC 1499 – The California Nebula**. Apparent dimensions 145′ × 40′, distance about 2,500 light-years.

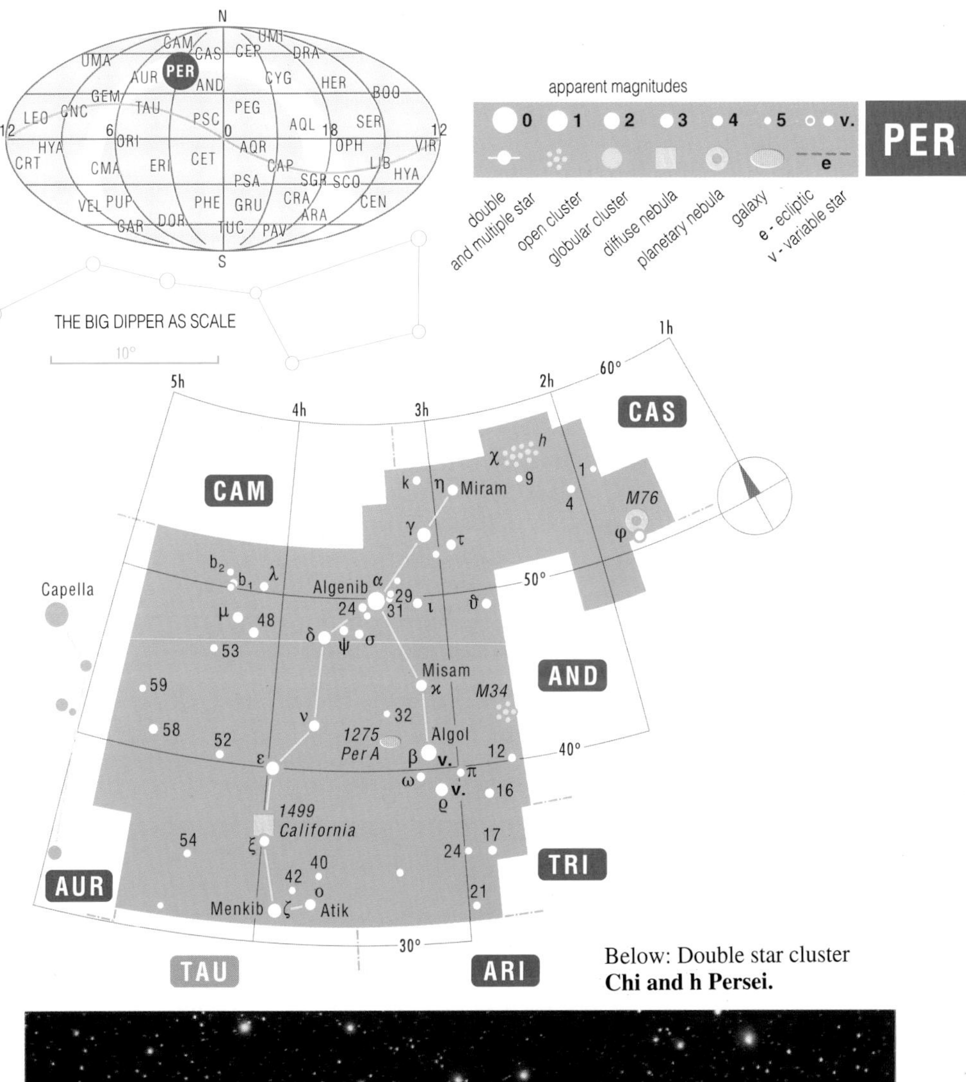

Below: Double star cluster **Chi and h Persei.**

PHOENIX

Phoenicis Phe The Phoenix

Phoenix, the fabulous bird, is the symbol of revival. According to mythology, the beautiful bird lived for 500 years. After its death it burned, and a new Phoenix rose from its ashes. The sacred bird was changed into a not very distinct constellation by Johann Bayer when he introduced 12 new southern constellations in his stellar atlas *Uranometria* in 1603. Phoenix can be found easily thanks to the bright Achernar in its proximity.

The brightest star is **Alpha Phoenicis**, a yellow giant of spectral class K0, situated at a distance of 77 light-years. The multiple systems of **Beta** and **Zeta Phoenicis** (opposite below) are objects for major telescopes.

Phoenix is situated in an area providing a vista of the world of distant galaxies. **NGC 87–89** and **NGC 92** (below) is a foursome of galaxies of considerably different types, situated at a distance of some 180 million light-years. NGC 87 is an irregular galaxy, NGC 88 a spiral galaxy with an outer gas envelope, NGC 89 a galaxy of Sa type with two distinct broad arms. The biggest of them is NGC 89, a type Sa galaxy with a single exceptional spiral arm about 100,000 light-years long.

Four galaxies **NGC 87–89** and **NGC 92**.

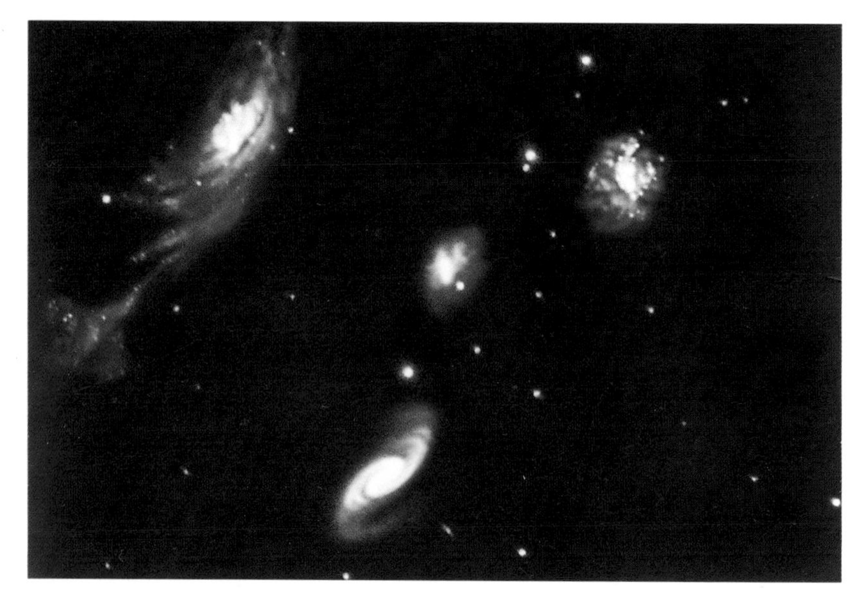

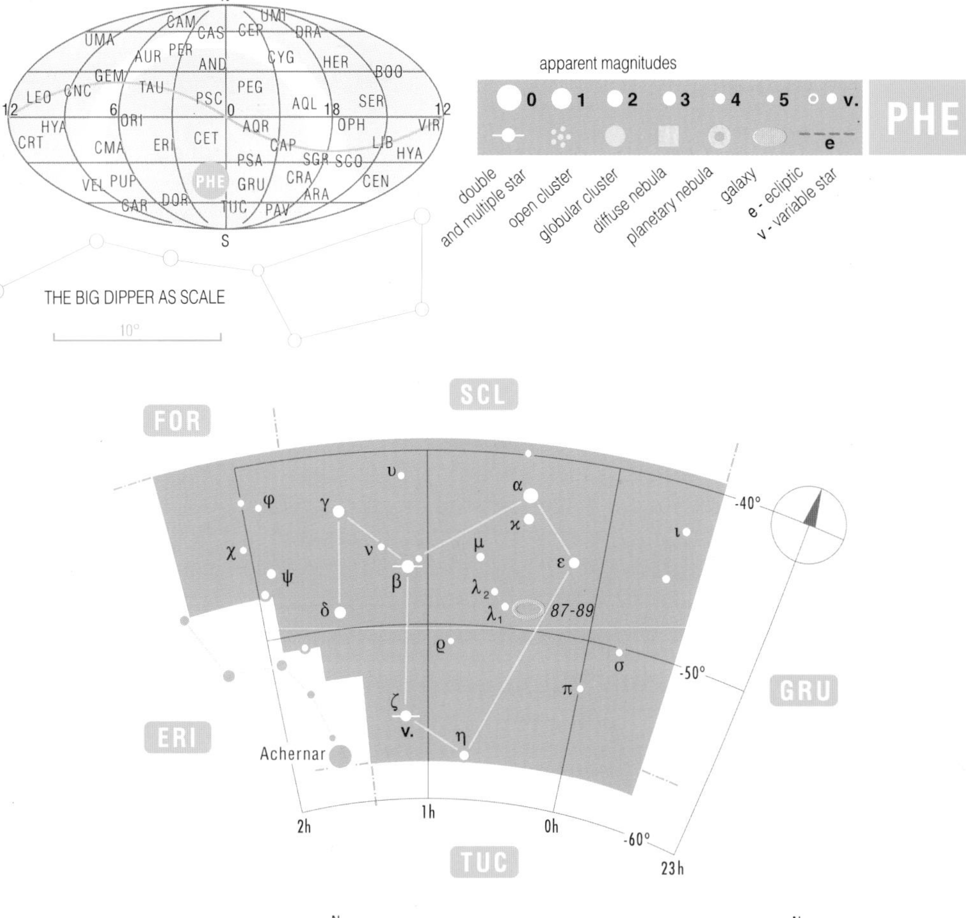

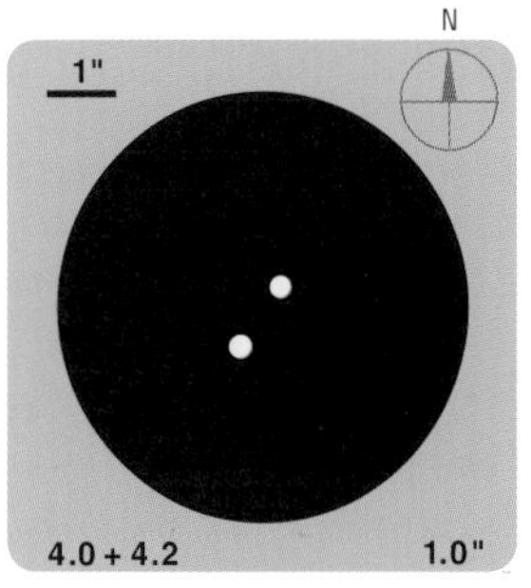

Beta Phoenicis is a multiple system at a distance of 198 light-years. Another companion of 11.5 mag is situated 58" from the main star.

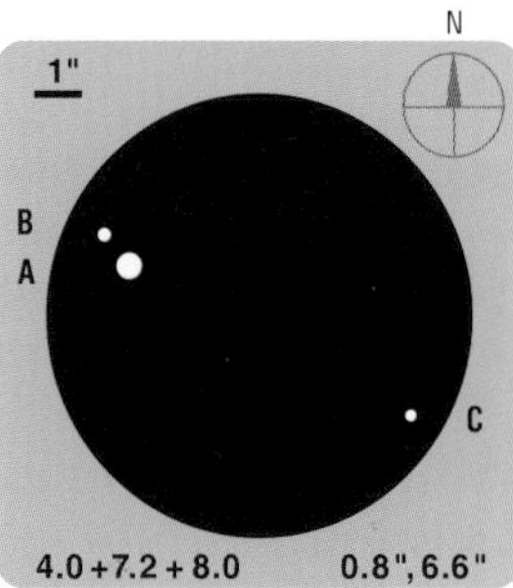

Multiple system **Zeta Phoenicis** at a distance of 280 light-years. Component A is an eclipsing binary.

PICTOR

Pictoris Pic The Painter

One of the 14 not very distinct constellations of the southern sky introduced by Nicolas-Louis de Lacaille in the middle of the 18th century. The original name was Equuleus Pictoris – the Painter's Easel, from which the present name Pictor – the Painter – has been derived. In some languages, however, the original name is still in use.

Alpha Pictoris, 3.3 mag, is a star of spectral class A7 at a distance of 99 light-years. Somewhat fainter but much more interesting is **Beta Pictoris**, 3.8 mag, of spectral class A5, at a distance of 63 light-years. In 1983, a disk of solid dust and ice particles around the star was discovered reaching to a distance of some 1,000 AU (astronomical units) from the star. It is possible that new planets are originating or have originated there.

Kapteyn's Star, 8.9 mag, is a red dwarf of spectral class M0, situated at a distance of merely 12.8 light-years, discovered in 1897 by Professor J. C. Kapteyn from the Netherlands. It is remarkable because of its fast motion, second only to Barnard's star (see Ophiuchus). It travels through space at the velocity of 280 km/sec.

Starry Sky in the Computer

To the users of personal computers, the software market offers innumerable astronomical programs. However, users should realize that the most expensive need not always be the best. A number of excellent, accurate programs can be obtained as freeware or shareware for a minimum fee. The beginning amateur astronomer will use primarily one of two software types: "astronomical yearbook" or "planetarium." Programs of the "yearbook" type provide all the data required for astronomical observations, such as times and positions of phenomena and objects, the times of rise and setting of celestial bodies, etc. The use of the program usually is quite simple and is based on the selection from the offered menu.

The programs of the "planetarium" or "stellar map" type make it possible to picture on the monitor any part of the starry sky with all objects that can be observed at a given time and on a given site with the naked eye or a telescope. The picture can be supplemented with the description of the objects, their coordinates and various data on selected objects. The computer may also store a whole library of astronomical tables, yearbooks, catalogues, maps and atlases. If you have a notebook, all these data can be readily available wherever you go.

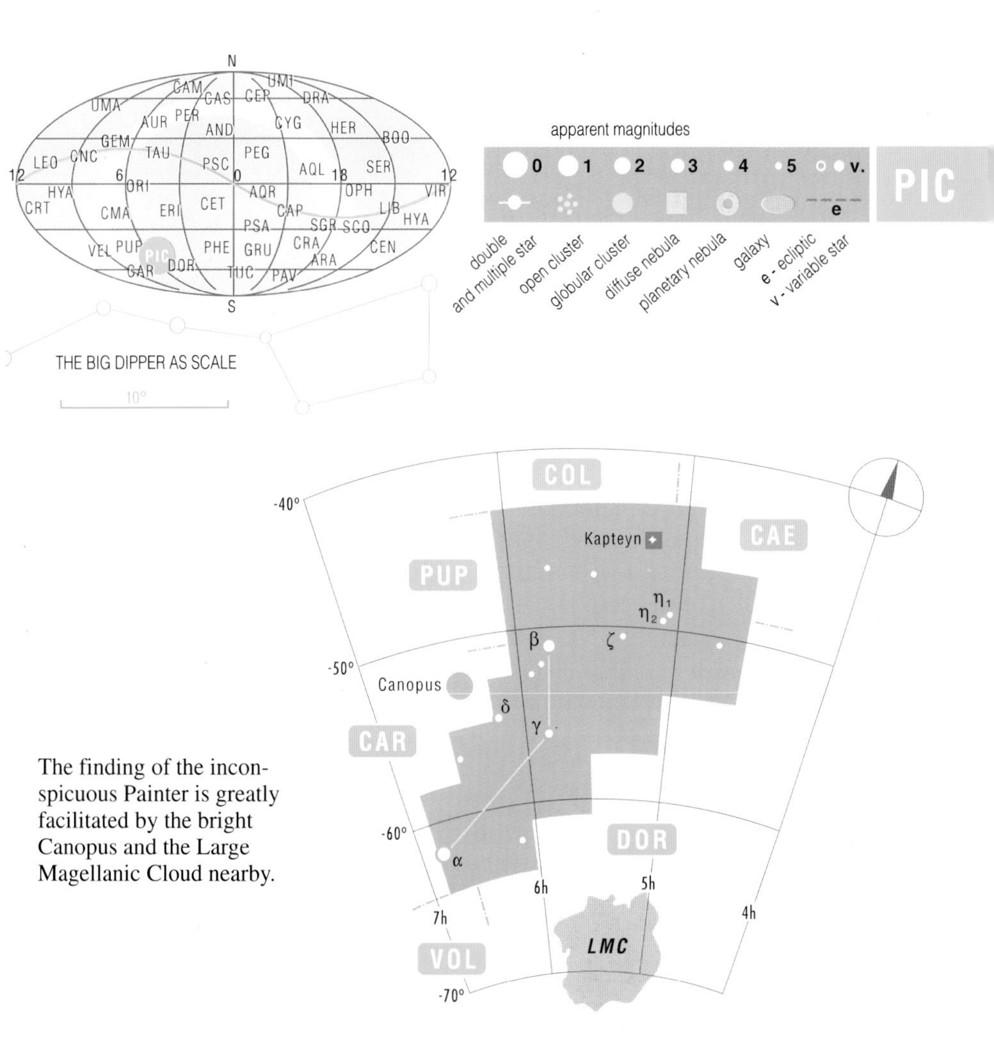

The finding of the inconspicuous Painter is greatly facilitated by the bright Canopus and the Large Magellanic Cloud nearby.

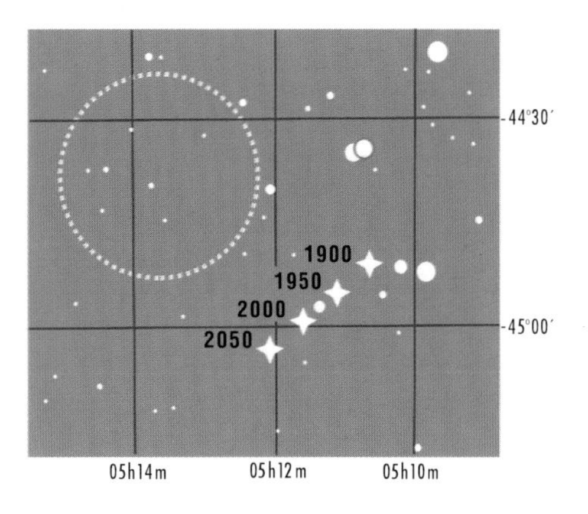

The map on the right shows the motion of **Kapteyn's Star** during the period of 1900 to 2050. The star travels an apparent distance of 8.7" per year. In two centuries, it travels in the sky 0.5°, one diameter of the full Moon (dotted circle).

PISCES

According to Greek (and Roman) mythology, the goddess Aphrodite and her son Eros (or Venus and her son Cupid) changed into fishes to escape the monster Typhon. We know the metamorphosis into a fish from the tale of the Capricorn (the Ram) which was only partly successful. The Fishes are a zodiacal constellation known for several millennia. On pictorial maps, it was shown as a pair of fishes with ribbons tied to their tales. Both ribbons were tied into a knot near Alpha Piscium – Al Risha (the Knot). The eastern fish, characterized by the stars Sigma to Phi Psc, can be found below Andromeda; the western fish is characterized by the Circlet consisting of the stars Iota, Theta, Gamma, Kappa and Lambda Psc.

Zeta Piscium is a good object for a small telescope. It has two components of 5.2 and 6.4 mag situated at a mutual apparent distance of 23". The distance of the components from the Sun is 148 and 195 light-years. Another pretty double star is **Psi 1 Piscium** with the components of 5.3 and 5.6 mag at a mutual distance of 30". **Van Maanen's Star**, 12.4 mag, which can be observed with a telescope with a 20 cm objective, is a white dwarf of about the Earth's size, but the Sun's mass, as a result, its density is about 1,000 kg/ccm. It is one of the smallest known stars, situated at a distance of merely 14.4 light-years.

Spiral galaxy **M 74 – NGC 628**, 9.5 mag, of Sc type, apparent diameter 10'. One of the faintest Messier's objects. Its distance is estimated within the limits of 25 and 42 million light-years.

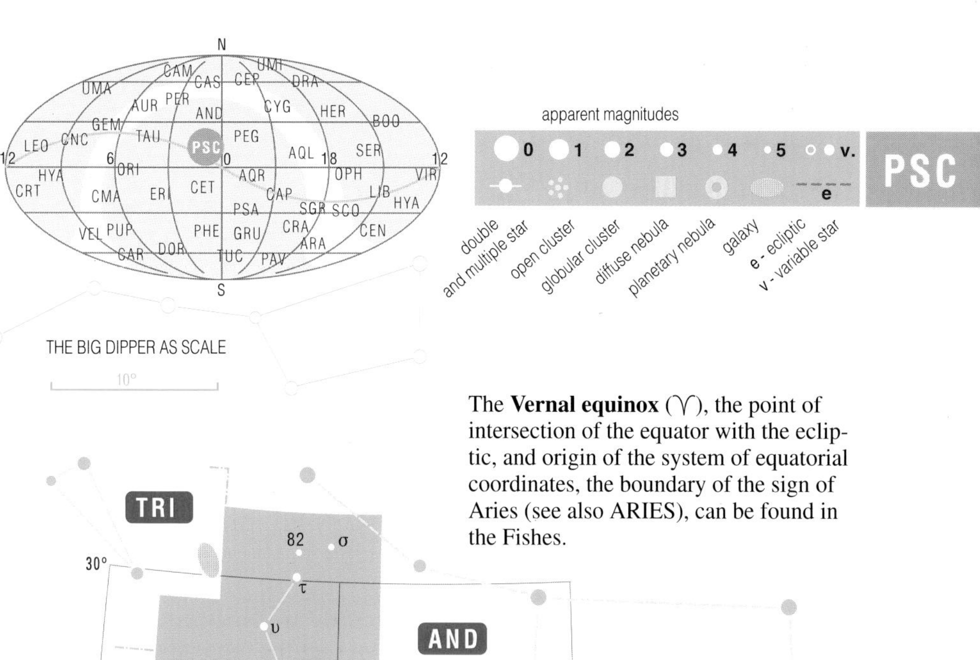

The **Vernal equinox** (♈), the point of intersection of the equator with the ecliptic, and origin of the system of equatorial coordinates, the boundary of the sign of Aries (see also ARIES), can be found in the Fishes.

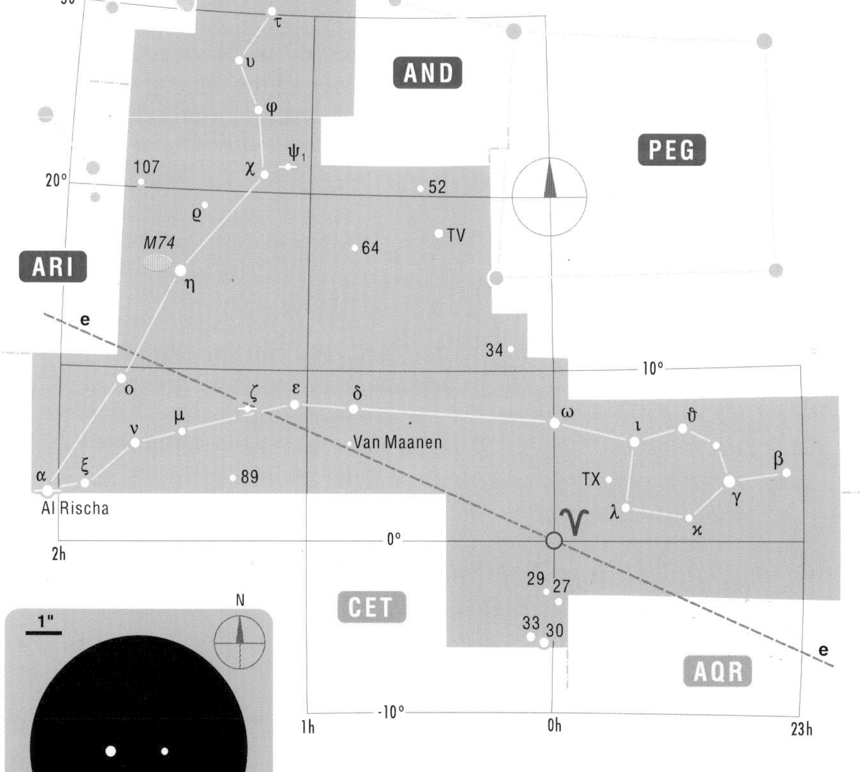

Left: Double star **Alpha Piscium – Al Risha**. Both components are spectroscopic binaries.

PISCIS AUSTRINUS

Piscis Austrini PsA The Southern Fish

The Southern Fish ranks among the oldest constellations known for millennia. It is adjacent to the Water Carrier, and old pictorial maps show it with the mouth in the stream of water flowing from the Water Carrier's jug. Its finding is facilitated by Fomalhaut, a solitary star of the first magnitude, one of the principal orientation points in southern sky.

Alpha Piscis Austrini – Fomalhaut, 1.2 mag, spectral class A3, is about twice as large as the Sun. It is situated relatively near, at a distance of merely 25 light-years from the Sun. In 1993, a cold ring of dust was discovered around Fomalhaut reaching to a distance of more than 400 astronomical units (AU) from the star. It is one of several similar discoveries of the 1990s, testifying to the fact that many stars are surrounded with gas and dust in which new planets are originating or perhaps have originated already.

The star **Lacaille 9352**, 7.4 mag, is a red dwarf of spectral class M2, with the fourth fastest proper motion known (see the map on the opposite page) after Barnard's Star in Oph, Kapteyn's Star in Pic and Groombridge 1830 star in UMa. Lacaille 9352 is situated at a distance of 10.7 light-years.

Photographing the Starry Sky

Simple photographs of the starry sky and even of whole constellations can be taken with a fixed camera on a tripod. You can use any camera with timer on the shutter that will let you set exposures from several seconds' to several minutes' duration. The picture will show the traces of the stars due to the rotation of the Earth. If you want to obtain pointlike pictures of stars, the camera must follow accurately the rotation of the sky during the whole exposure. This can be achieved with a so-called equatorial mounting, standard on astronomical telescopes. The ISO rating of the films should be as high as possible. Modern color films generally work well.

Detailed pictures of celestial objects are made by means of a telescope serving as a big teleobjective. The eyepiece is replaced by the camera body fastened to the telescope by means of an adapter. A single-lens-reflex camera with exchangeable objectives is best suited for this purpose. The sensitive film is then directly in the focus of the telescope objective.

Astrophotography offers enormous possibilities to amateurs who can use even electronic picture recording by means of CCD cameras and subsequent computer image processing.

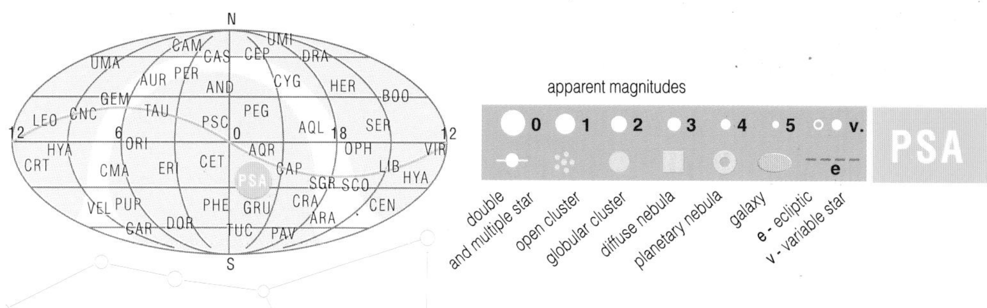

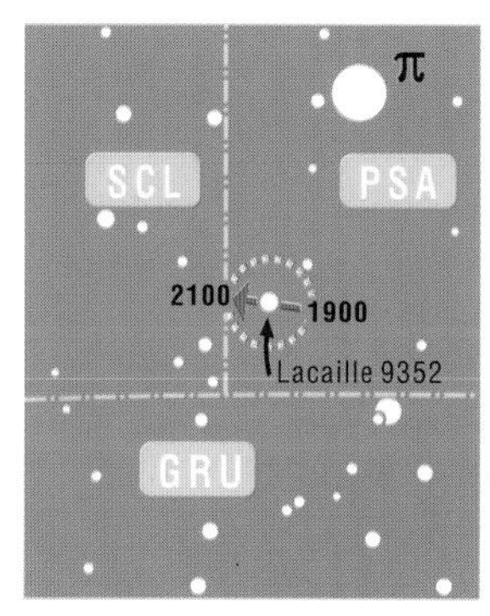

CET
CAP
-20°
ε
λ
η
α
Fomalhaut
ϑ
-30°
SCL
τ
μ
δ
γ
β
ι
υ
π
MIC
Lacaille 9352
23 h
22h
21 h
-40°
GRU

Above: The motion of **Lacaille 9352** from 1900 to 2100, when it will enter the Sculptor constellation. The dotted circle represents the size of the full Moon, 0.5°. The star traverses 6.9" in one year.

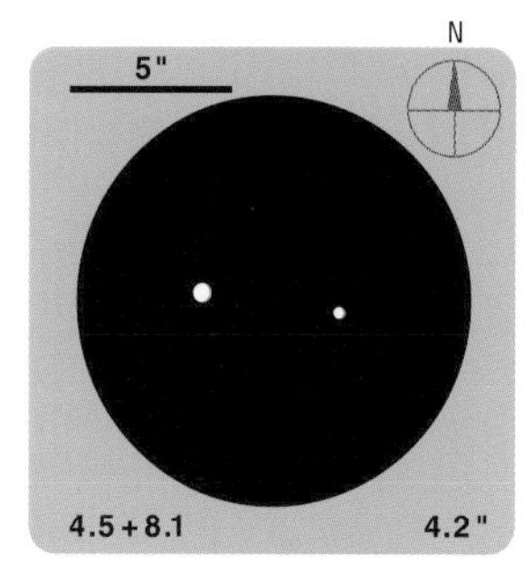

Left: Double star **Gamma Piscis Austrini** is situated at a distance of some 220 light-years. Its brighter component is of spectral class A0.

Right: **Beta Piscis Austrini** is an optical double which can be discerned even with a small telescope.

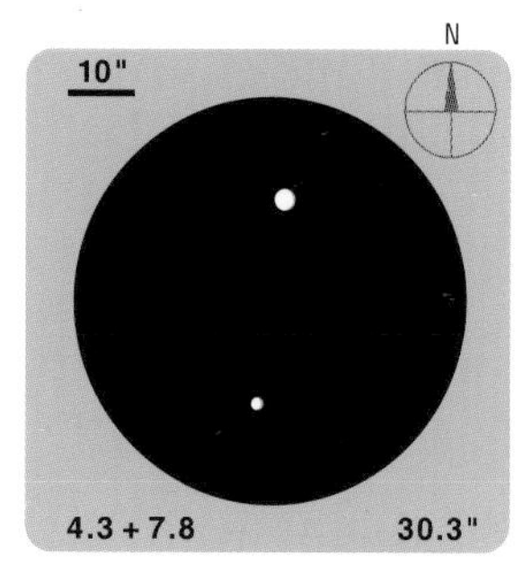

PUPPIS

Puppis Pup — The Stern

The number of classical constellations of ancient times also includes the mythological Ship Argo (Argo Navis) on which the Argonauts made their famous voyage in quest of the Golden Fleece. After the final division of the sky, the ship was divided into four separate constellations proposed by Lacaille: Puppis (the Stern), Carina (the Keel), Vela (the Sails) and Pyxis (the Mariner's Compass). Together they number almost one thousand stars visible to the naked eye.

The starry sky lovers will be attracted by a great number of bright open star clusters visible with binoculars. **M 47 – NGC 2422** can be observed even with the naked eye. Within a 25' circle there are about 50 stars of 6.0 to 12.0 mag. The star cluster **NGC 2451** with an apparent diameter of 45' contains about 40 stars grouped around the star **c Puppis**. **M 46 – NGC 2437** is a dense open star cluster of circular outline and apparent diameter of 20'. It contains some 100 stars of 9.0 – 13.0 mag, situated at a distance of 5,400 light-years.

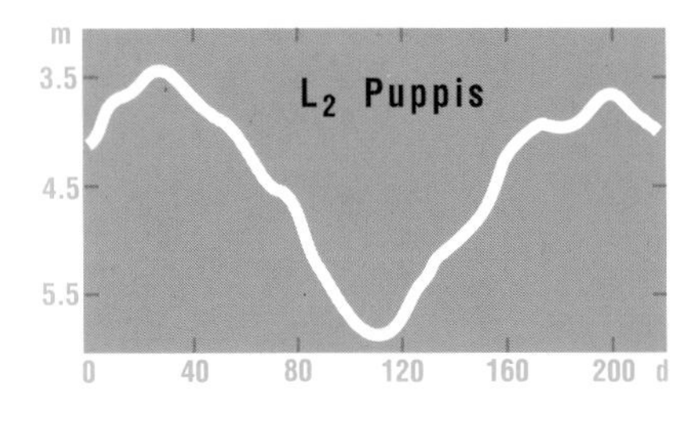

The light curve of the long-period variable star **L_2 Puppis**. This red giant of spectral class M5 pulsates and changes brightness within 141 days.

PYXIS

Pyxidis Pyx — The Mariner's Compass

Originally called Pyxis Nautica, the Nautical Compass. Nicolas-Louis de Lacaille placed this important navigation instrument in the area where the mast of the Argo used to be. Consequently, it is not a part of the original equipment of the ship, but one of the scientific instruments which Lacaille immortalized in the southern sky.

Beta Pyxidis, 4.0 mag, is a yellow supergiant of spectral class G5 with a companion of 12.5 mag at an apparent distance of 12.6". The star **T Pyxidis** is a so-called recurrent nova. Its outbursts repeated after 12 to 25 years increase its brightness from the minimum of 15 mag to the maximum of some 7.0 mag.

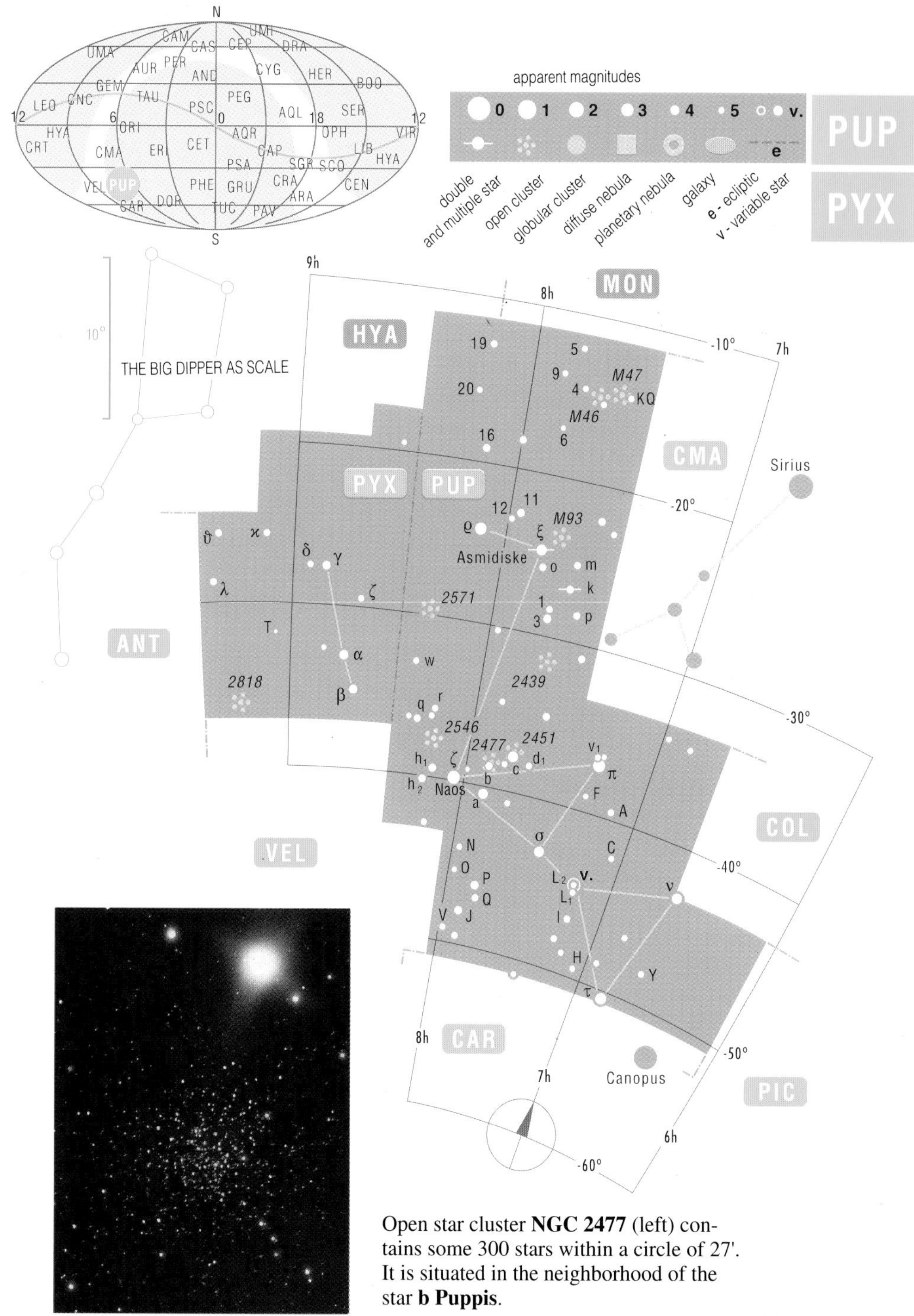

Open star cluster **NGC 2477** (left) contains some 300 stars within a circle of 27'. It is situated in the neighborhood of the star **b Puppis**.

RETICULUM

Reticuli Ret The Reticle

The number of scientific instruments and aids memorialized by Nicolas-Louis de Lacaille in the sky also includes Reticulum (the Reticle). Its original name was Reticulus Rhomboidalis, the rhomboidal reticle. It should recall the reticle or net of fine fibers in the field of view of a telescope facilitating astronomical observation. Lacaille himself made numerous measurements of star positions in the southern sky from 1750 to 1754, when he headed an expedition of the French Academy of Sciences to the Cape of Good Hope. This minute constellation can be found in the vicinity of the Large Magellanic Cloud (LMC).

The three brightest stars of the Reticle are double stars with faint companions. **Alpha Reticuli**, 3.4 mag, has a companion of 12.0 mag at an apparent distance of 48.5". The distance of the star is 163 light-years. **Beta Reticuli**, 3.8 mag, with a companion of 8.0 mag, is situated at a distance of 100 light-years from the Sun. Also **Epsilon Reticuli**, 4.4 mag, is a double star, situated 60 light-years away. Its companion of 12.5 mag is 13.7" distant from the principal component. The optical double star **Zeta 1** and **Zeta 2** (components of 5.2 and 5.5 mag) can be separated with the naked eye.

Observation Records

Every observation is valuable and irreplaceable for the individual observer. However, it can also have broader significance. Amateur astronomers have a great many opportunities for valuable professional observations even today. To make this activity useful, it is necessary to keep accurate and systematic records necessary for subsequent evaluation.

The first information to record is the precise time of observation. It must be clear what zone time it is – beware of daylight savings time! The time frequently used in astronomy is the ***Universal Time*** *(UT), the time at the Greenwich meridian. Every observer should know the difference between his zone time and the Universal Time. The record should also contain the observer's geographical position, the instruments used, and sometimes also meteorological and other data, depending on the object and purpose of observation.*

Numerous, perhaps most UFO's (Unidentified Flying Objects) originated because the observation reports were inadequate. Do not produce further UFO's!

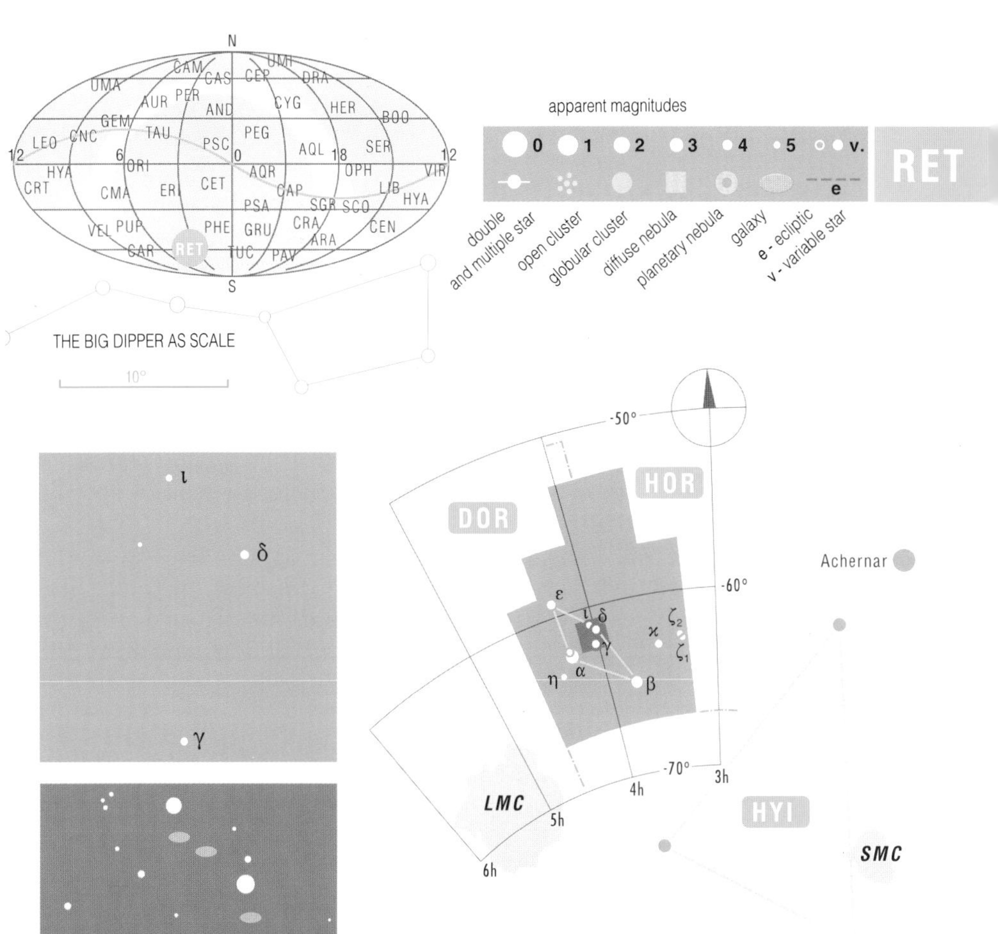

The three pictures on the left clearly show the increase in the number of stars seen if we proceed to fainter and still fainter magnitudes. The upper map shows a square with 1.25°-long sides (about 2.5 times the diameter of the Moon) containing the stars Iota, Delta and Gamma Reticuli and showing the stars of up to 7.0 mag. The center map shows the same square and the stars and galaxies of up to 11.0 mag. The lowest map shows the same square with stars and galaxies of up to 15.0 mag. Big telescopes make it possible to see even substantially fainter objects. How little we can see with the naked eye!

SAGITTA

Sagittae Sge The Arrow

This tiny, but characteristic and easy-to-identify constellation was known under the same name to astronomers of different ancient nations. With its area of 80 square degrees it is the third smallest constellation, larger only than the Southern Cross (Crux) and the Colt (Equuleus). The Arrow was connected with many different myths: according to one of them it was the arrow with which Apollo killed the one-eyed giant, Cyclops. According to another version the arrow was released by Cupid, the god of love.

Alpha Sagittae, 4.4 mag, is a yellow giant of spectral class G0. It has four companions, mostly fainter than 11.0 mag. The brightest star of the constellation is **Gamma Sagittae**, 3.5 mag, an orange giant of spectral class K5, shining from some 275 light-years.

M 71 – NGC 6838 (opposite below) is a loose globular star cluster without the usual high star concentration in its nucleus. Therefore, it is described sometimes as a rich open star cluster. Its distance from the center of the Galaxy is about 24,000 light-years, that from the Sun 14,500 light-years.

M 17 – NGC 6618 – The Omega Nebula (to p. 182).

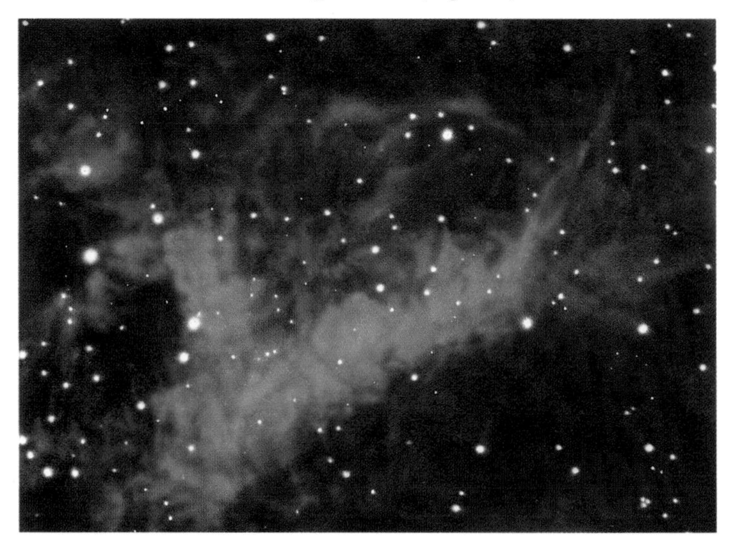

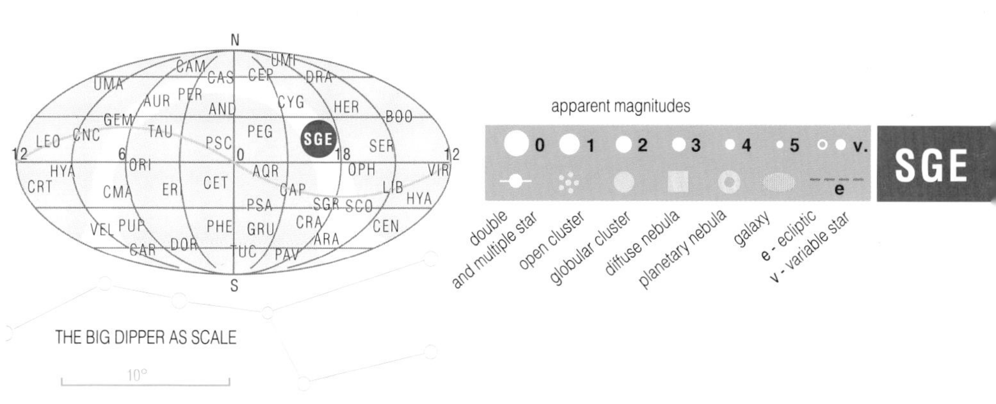

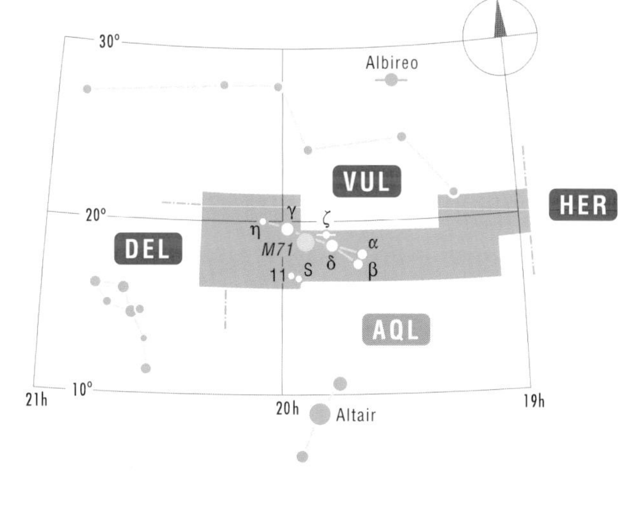

Below: Double star **Zeta Sagittae** is a physical pair with joint movement of its components through space. Its distance from the Sun is about 325 light-years.

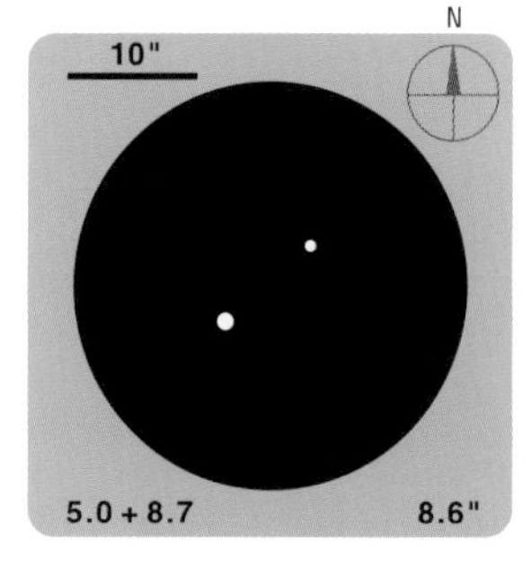

Right: Loose globular star cluster **M 71 – NGC 6838**, total brightness 8.0 mag, apparent diameter 6'. Its brightest stars are of 12.0 mag. The object can be observed with binoculars as a hazy spot.

SAGITTARIUS

Sagittarii Sgr The Archer

Zodiacal constellation shown on pictorial maps as the mythical Centaur – half man, half horse – with a bow and an arrow pointing at the adjacent Scorpion. On modern maps it is usually represented by the drawing of a "teapot."

The stars of the Archer can be observed in the brightest parts of the Milky Way in the direction of the center of the Galaxy. It is a constellation with the most abundant number of Messier's object visible with binoculars. Their number includes the diffuse nebulae **M 8 – The Lagoon Nebula**, **M 20 – The Trifid Nebula**, and **M 17 – The Omega**, or Horseshoe, **Nebula**, open star clusters **M 23**, **M 25**, and **M 21**, and the star cloud **M 24**. The globular star cluster **M 22 – NGC 6656**, the richest of the numerous globular star clusters in the Archer, has an apparent diameter of 17'. At a distance of 10,000 light-years it is one of the nearest objects of its type.

The nebula **M 17**, called **Omega**, or **Horseshoe** (see p. 180), of apparent dimensions of 45' × 35', is situated at a distance of some 6,000 light-years. Equally distant is also **M 8 – NGC 6523 – Lagoon** (below) surrounding the loose star cluster **NGC 6530** consisting of very hot, young stars.

Diffuse nebula **M 8 – Lagoon**.

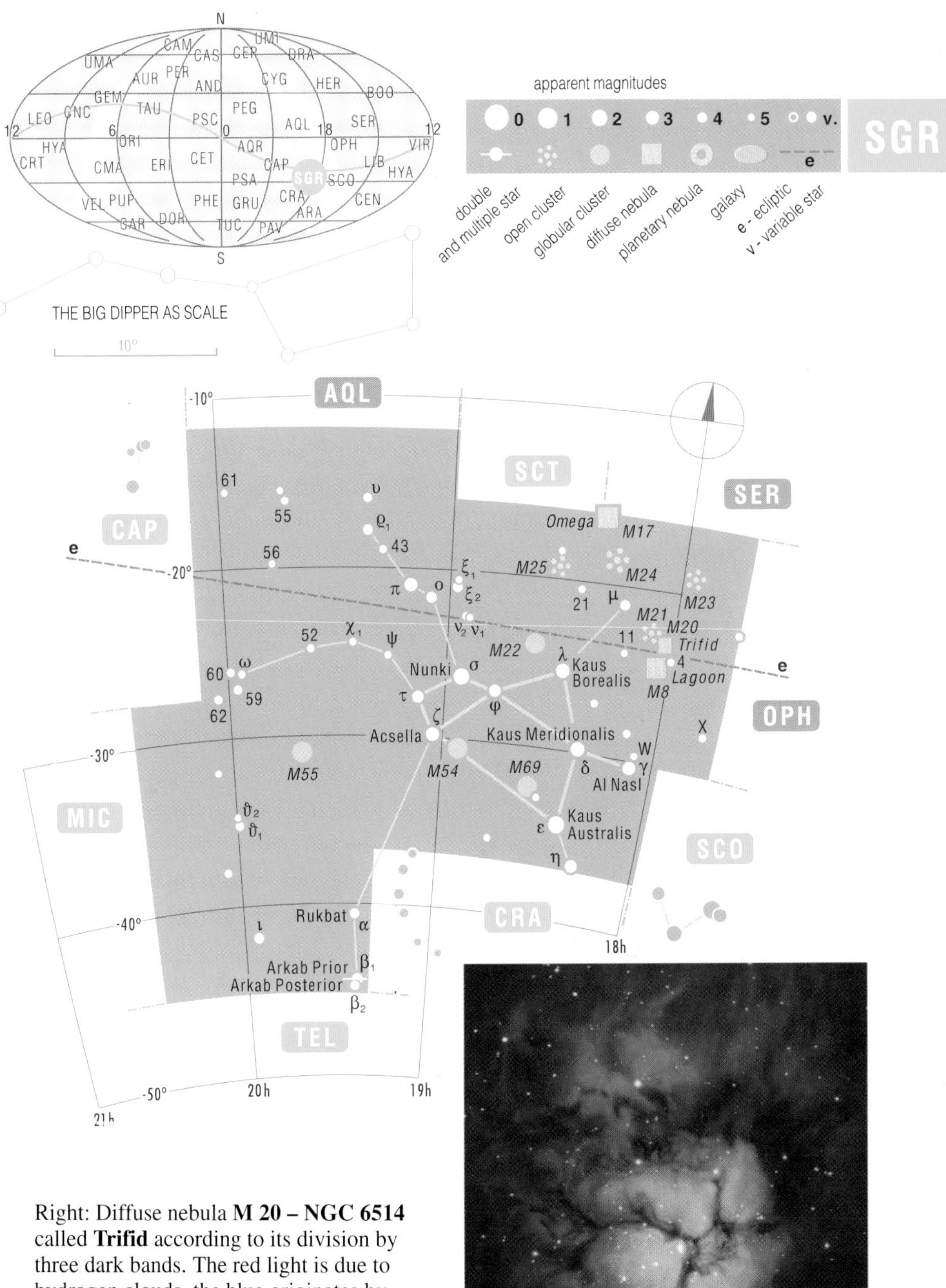

Right: Diffuse nebula **M 20 – NGC 6514** called **Trifid** according to its division by three dark bands. The red light is due to hydrogen clouds, the blue originates by the diffusion of star light in the dust particles of the nebula.

SCORPIUS

Scorpii Sco The Scorpion

This zodiacal constellation can be seen in all its beauty from observation sites situated south of the 40th parallel of North latitude. It is one of the few constellations the shape of which corresponds to its name. In the sky, it recalls the dangerous animal which, according to ancient myths, killed Orion the hunter. The original Scorpion had powerful pincers reaching as far as the present-day Scales (Libra).

Alpha Scorpii – Antares ("Against-Mars," or the Rival of Mars, so called because of its red color) is a red supergiant of spectral class M1 which changes its brightness irregularly from 0.9 to 1.1 mag. Its diameter is about 950 million km, its mass only 15–25 suns, as a result of which the density of the giant is extremely low. Should it be in the place of the Sun, the star would reach beyond the orbit of Mars.

The open star cluster **M 7 – NGC 6475**, total brightness 3.3 mag, provides a wonderful view in binoculars and a small telescope. In an 80' circle, it is possible to see 50 to 80 stars, the brightest of which are of 7.0 mag. The distance from the Sun is about 800 light-years. It is the most southern Messier's object.

The environs of **Antares** (lower left). The globular star cluster **M 4 – NGC 6121** to the right from Alpha Scorpii can be observed with binoculars, but its true beauty is revealed only by the telescope. It is one of the brightest, biggest (apparent diameter 26') and nearest globular star clusters, situated at a distance of some 7,000 light-years. The bright star Sigma Scorpii is above M 4, on the right. The blue nebula on top surrounds the star Rho Ophiuchi.

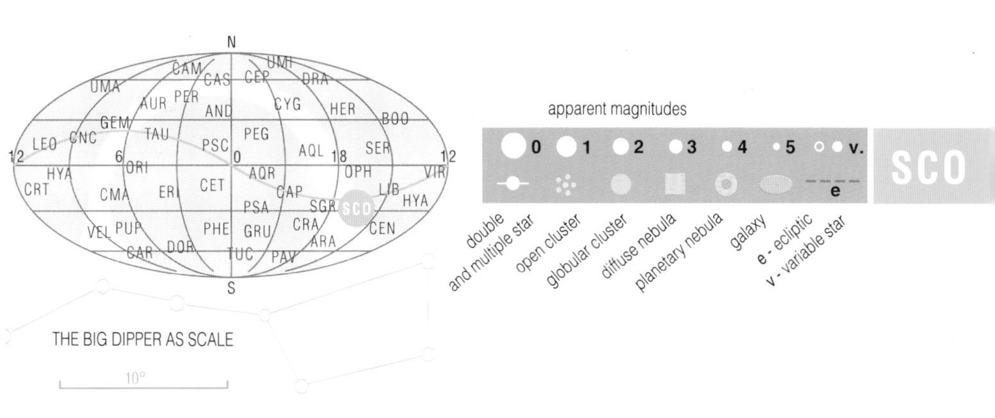

Below: **Beta Scorpii** is a multiple system: in a small telescope it is a pretty double star both components of which are spectroscopic binaries.

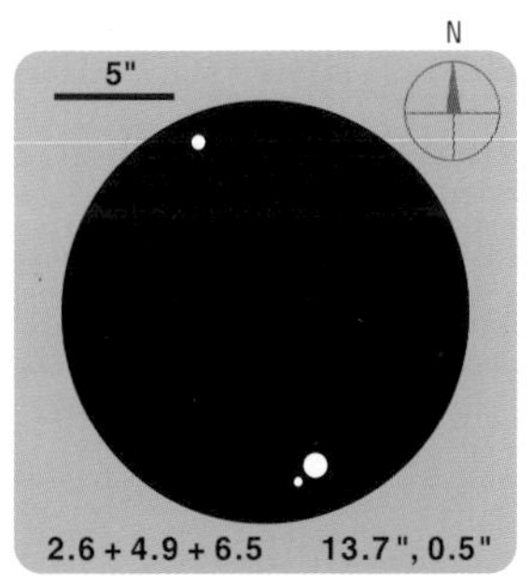

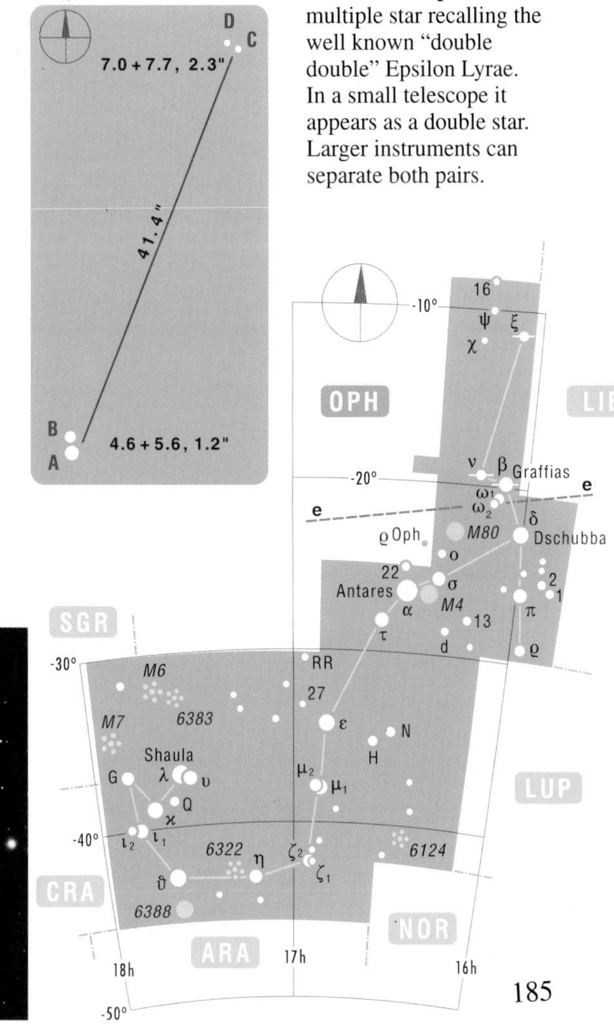

Left: **Nu Scorpii** is a multiple star recalling the well known "double double" Epsilon Lyrae. In a small telescope it appears as a double star. Larger instruments can separate both pairs.

Below: **M 6 – NGC 6405**, one of the most beautiful open star clusters observed with a small telescope. Its outline earned it the name of **Butterfly Cluster**.

SCULPTOR

Sculptoris Scl The Sculptor

The Sculptor is also one of the "artifacts" of Nicolas-Louis de Lacaille, although its initial name was broader: L'Atelier du Sculpteur – Sculptor's Studio. However, do what we may, we cannot see either the studio or its master, as this barren celestial space contains only 6 stars brighter than 5.0 mag. Nevertheless, it also contains the South Galactic Pole (S.G.P. on the map) which means that we are looking in the direction perpendicular to the plane of our Galaxy into the most remote parts of the universe.

The dwarf galaxy called the **Sculptor System** or **Sculptor dE** (below) is a member of the Local Group of galaxies containing about two dozen such "dwarfs" as well as three big galaxies. The next nearest group of galaxies is situated in the Sculptor, at a distance of some 10 million light-years. The two galaxies shown on the opposite page are also members of this group.

We observe **NGC 55**, total brightness 8.0 mag, from the side (opposite below). Most probably it is an irregular galaxy alike the Magellanic Clouds. **NGC 253**, a big spiral galaxy of Sc type (total brightness 7.0 mag, apparent dimensions 25′ × 7′), can be observed with binoculars. It ranks among the brightest galaxies in the sky.

The **Sculptor System** is a dwarf galaxy, but its appearance recalls a loose globular star cluster rather than a galaxy. Its distance is 260,000 light-years, its diameter about 8,000 light-years. Its brightest stars are only of 18.0 mag.

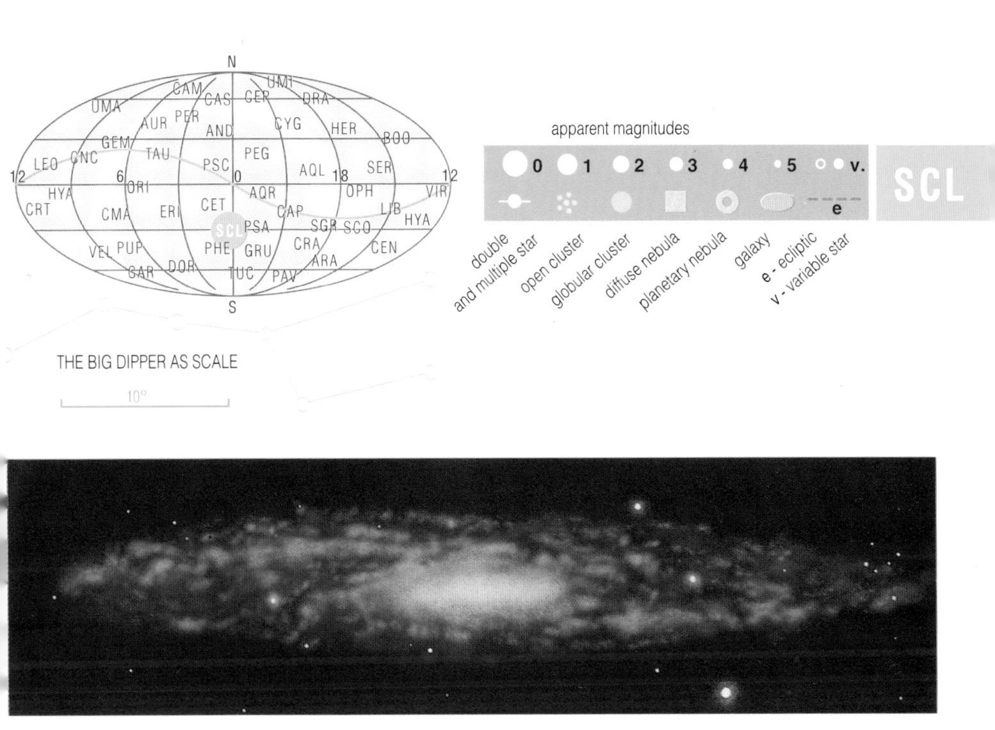

Spiral galaxy **NGC 253** (above).

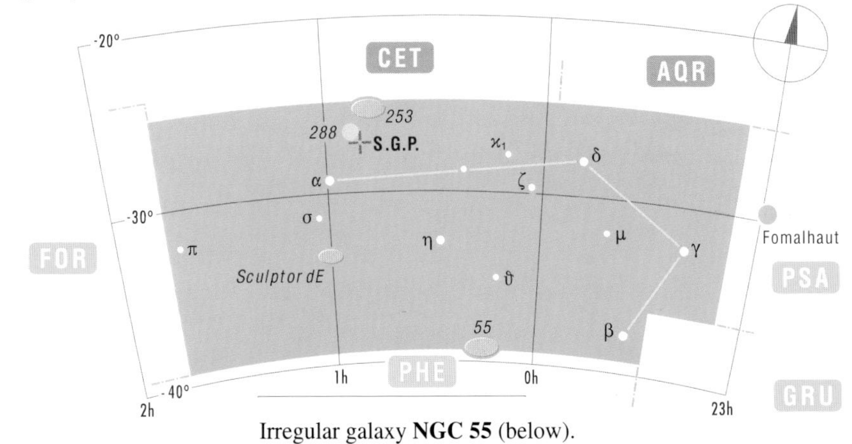

Irregular galaxy **NGC 55** (below).

SCUTUM

Scuti Sct The Shield

The Shield is a miniature constellation bordered by an austere rectangle. It was introduced in 1690 by Johann Hevelius, astronomer from Gdansk (Danzig) in honor of his king, John III Sobieski. The initial name of the constellation was Scutum Sobiescianum, "Sobieski's shield."

Alpha Scuti, 3.8 mag, the brightest star of the constellation, is an orange giant at a distance of 175 light-years. **R Scuti** is visible to the naked eye only rarely. It is an irregularly pulsating variable star, the brightness of which varies between 4.4 and 5.0 mag in the maximum and 6.0 and 8.7 mag in the minimum within a period of some 5 months. The considerable brightness variations may be due to carbon particles (smoke) originating in the atmosphere of the star.

The most attractive object in the Shield is the open star cluster **M 11 – NGC 6705**, one of the richest and brightest clusters (opposite below). It was discovered by G. Kirch, a German astronomer, in 1681. It contains one star of 8.0 mag, about 500 stars brighter than 14.0 mag and a rich group of faint stars of 11.0 to 14.0 mag.

The Shield contains also one of the brightest star clouds in the Milky Way. With the naked eye you will discover it as a brighter spot in the band of the Milky Way. Details can be appreciated on a photograph only.

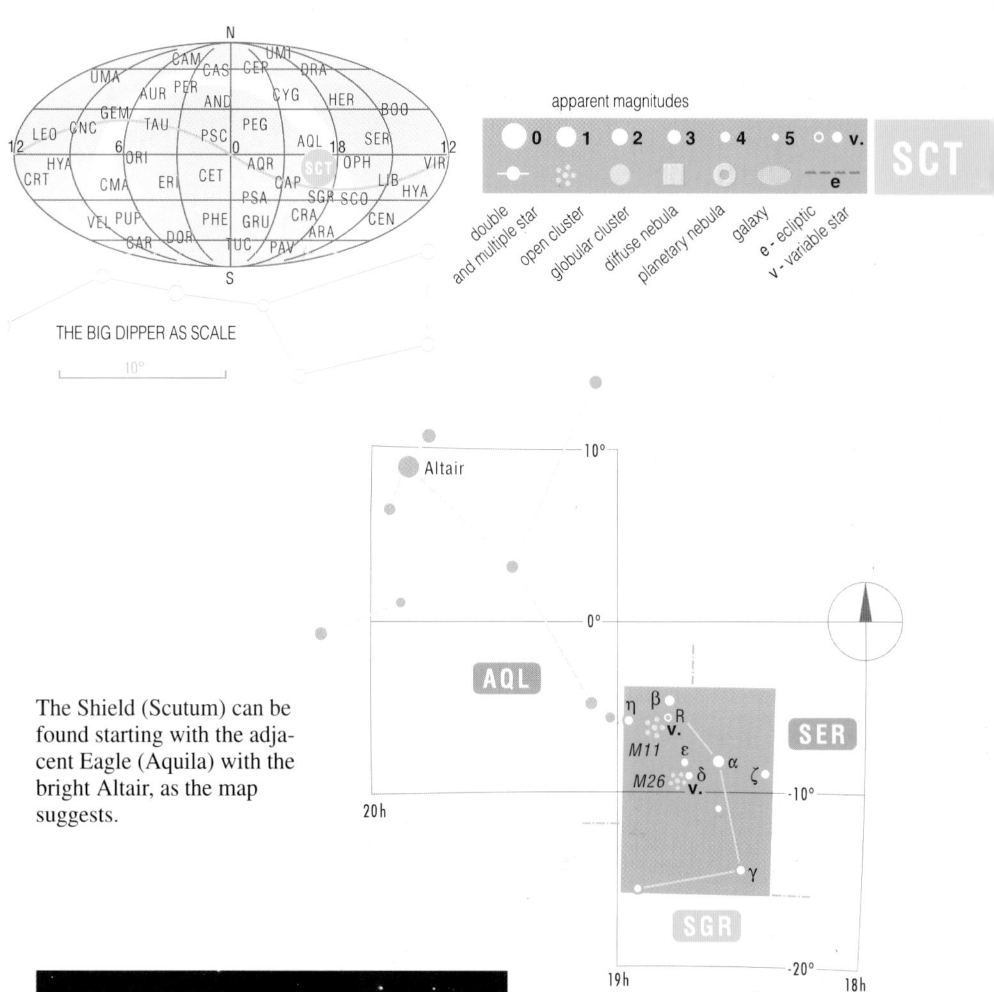

The Shield (Scutum) can be found starting with the adjacent Eagle (Aquila) with the bright Altair, as the map suggests.

The open star cluster **M 11** (left) has a total brightness of 6.0 mag and an apparent diameter of 13'. Its actual diameter is estimated at 15 to 20 light-years. The system contains about 500 stars brighter than 14.0 mag. Its distance from the Sun is about 5,500 light-years.

SERPENS

Serpentis Ser — The Serpent (Head & Tail)

The Serpent is the only constellation that is in two separate parts. Initially, the whole Serpent was connected with the Serpent Bearer (Ophiuchus – see Oph) to form one of the biggest and oldest constellations. The official constellation boundaries provided by IAU in 1930, however, have separated the **Serpent Head (Serpens Caput)** from the **Serpent Tail (Serpens Cauda)**, which reaches into the Milky Way. The central part of the serpent's body remained a part of the Serpent Bearer (Ophiuchus). It is a curiosity that can be explained by a look at some ancient pictorial sky maps.

Alpha Serpentis – Unuk Elhaia, Unukalhai, 2.6 mag, is an orange giant at a distance of 73 light-years. **Delta Serpentis** can be separated with a telescope as a double star with components of 3.8 and 4.9 mag at a mutual apparent distance of 4.0".

M 16 – NGC 6611 (see p. 192) is a large open star cluster with a diffuse nebula, one of the most beautiful objects of its type. In literature it is known as the Eagle Nebula or the Star Queen Nebula. The stars of the open cluster are hot giants of spectral classes O and B, born less than one million years ago.

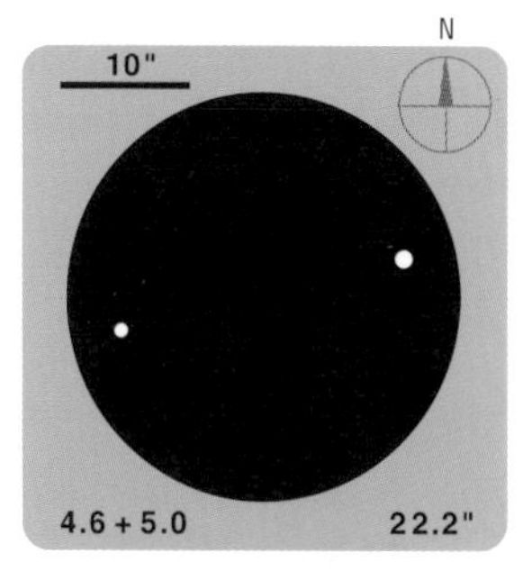

Theta Serpentis (above), a very pretty double star visible with a small telescope at a distance of some 110 light-years.

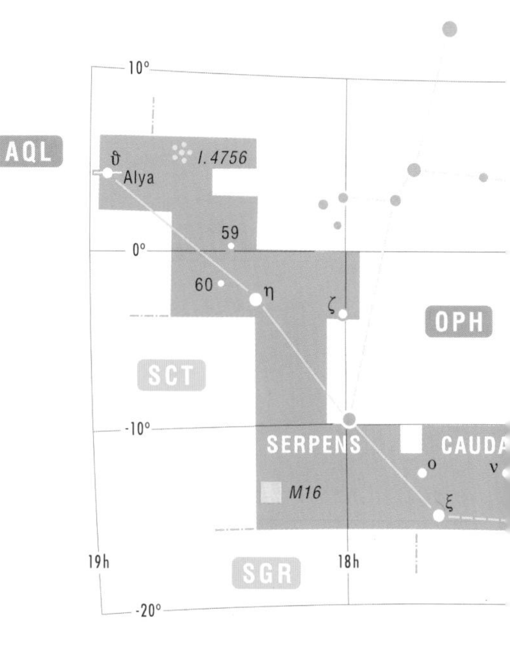

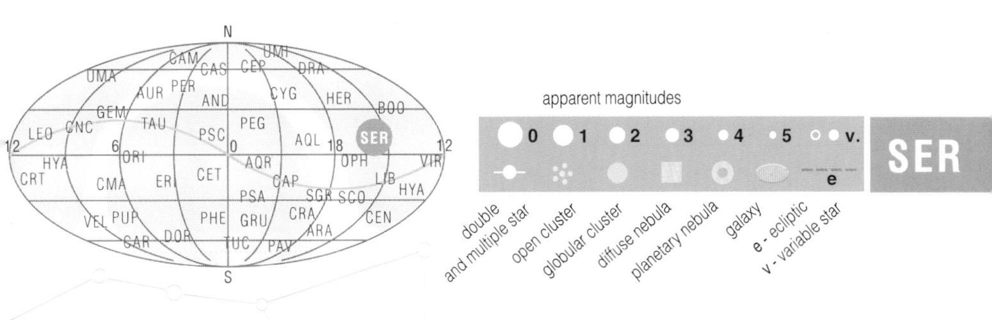

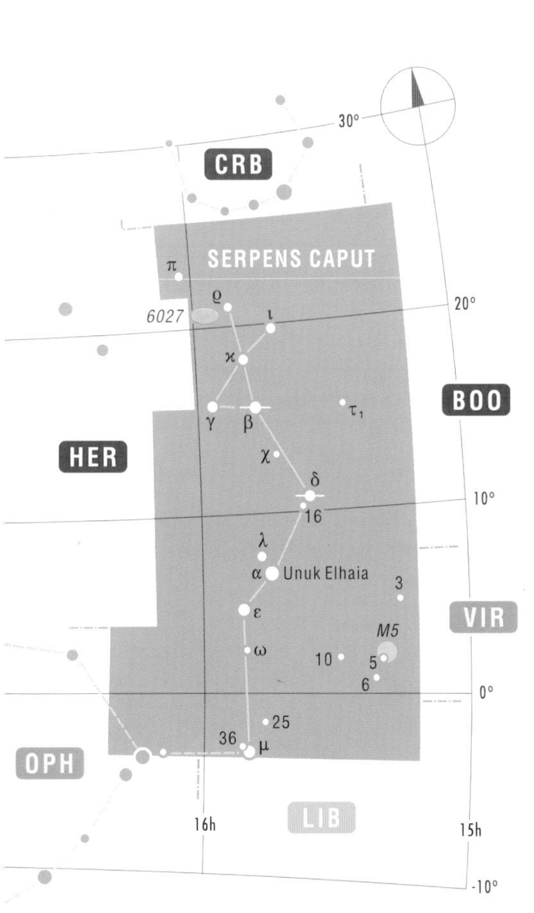

Right: Globular star cluster **M 5 – NGC 5904**, of total brightness of 5.7 mag and apparent diameter of 20′. Its distance from the Sun is 25,000 light-years.

Above: Group of galaxies around **NGC 6027** mutually linked by gravitational forces.

SEXTANS

Sextantis Sex The Sextant

The constellation was introduced by Johannes Hevelius at the end of the 17th century. Its original name was "Sextans Uraniae," the Sextant of Urania – the Muse of astronomy. The sextant of Hevelius was a big instrument on a tripod, used for the measurements of the angles or positions of the stars. Although it was not provided with a telescope yet and could be aimed at a star only by a combination of simple slits and the astronomer's sharp eye, the most experienced observers could attain an accuracy of nearly one arc minute.

The Sextant can be found only under good observation conditions, as it contains only two stars brighter than 5.0 mag. When looking for it in the sky you will be assisted by the nearby Regulus in Leo and Alpha Hydrae. The brightest star of the constellation is **Alpha Sextantis**, 4.5 mag, spectral class A0, situated at a distance of 287 light-years.

The galaxy **NGC 3115**, 9.0 mag, is within reach of small telescopes and its appearance is similar in the telescope and in the photograph (opposite below): it looks like a lens or a spindle – hence its name the **Spindle Galaxy**.

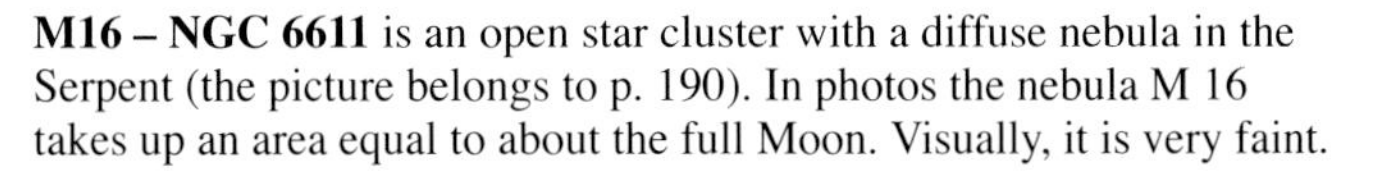

M16 – NGC 6611 is an open star cluster with a diffuse nebula in the Serpent (the picture belongs to p. 190). In photos the nebula M 16 takes up an area equal to about the full Moon. Visually, it is very faint.

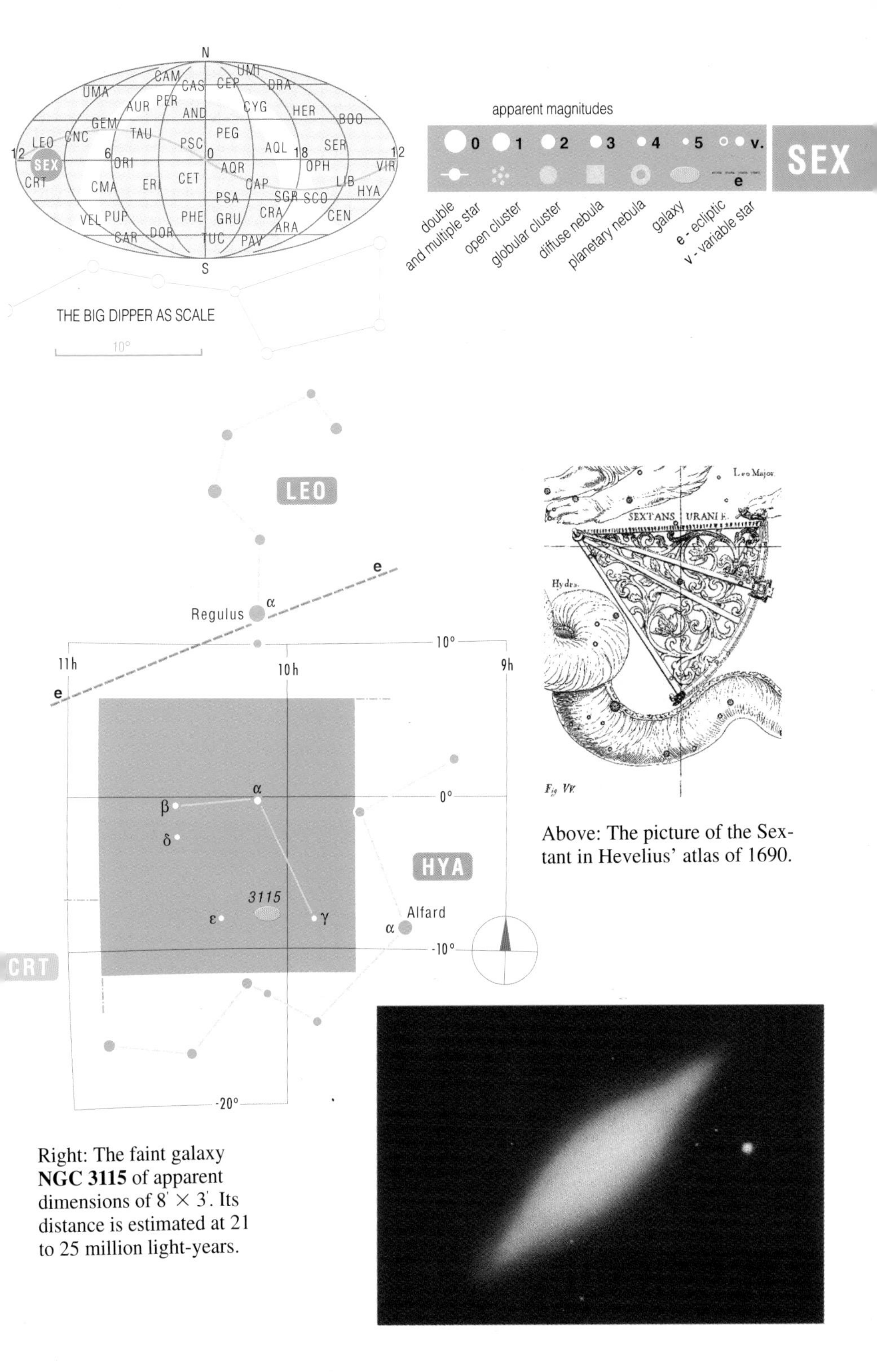

Above: The picture of the Sextant in Hevelius' atlas of 1690.

Right: The faint galaxy **NGC 3115** of apparent dimensions of 8′ × 3′. Its distance is estimated at 21 to 25 million light-years.

TAURUS

Tauri Tau The Bull

An ancient zodiacal constellation known to the oldest cultures. On pictorial maps Orion, the big hunter, is shown struggling with the Bull. The V-shaped outline of the Hyades group of stars indicate the bull's head, its eye shining with the orange Aldebaran, and its horns reaching to Beta and Zeta Tauri.

Alpha Tauri – Aldebaran, 1.0 mag, is an orange giant of spectral class K5 and a diameter of 36 Suns. It is situated at a distance of 65 light-years, far nearer than the Hyades to which it does not belong.

The **Hyades** (Melotte 25) are a so-called moving cluster: all of its stars travel through space along parallel trajectories. It has about 200 members. Its actual diameter is 8 light-years, its distance 151 light-years. Also **M 45 – The Pleiades** – are a moving cluster. 6–7 Pleiades can be observed with the naked eye, while the telescope shows about 100 of them. The diameter of the cluster is 7 light-years; its distance is 380 light-years.

M 1 – NGC 1952 – The Crab Nebula (picture on p. 196) is a remnant of a supernova which flared at a distance of 6,500 light-years in 1054. The dimensions of the nebula are 9 × 14 light-years, but the nebula continues to expand. The remnant of the original star is a so-called pulsar, a collapsed neutron star of about 10 km diameter and a mass of 1.4 sun rotating about its axis at 30 revolutions per second.

The Pleiades – an open star cluster.

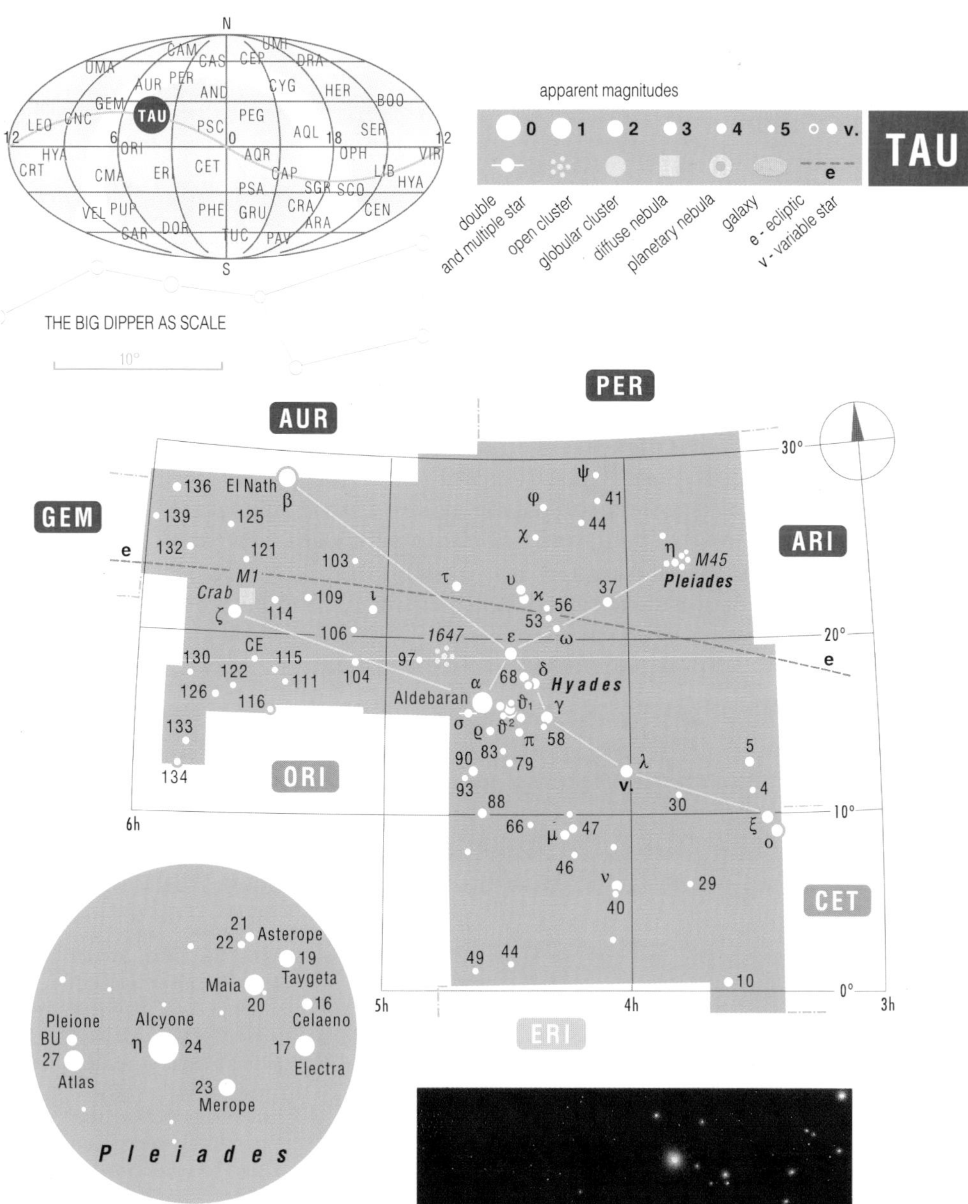

Above: The map of the **Pleiades** with star names and Flamsteed numbers. Right: The **Hyades** with the yellow-orange Aldebaran. Apparent diameter of the cluster is 3.5°.

TELESCOPIUM

Telescopii Tel The Telescope

The constellation was introduced by Nicolas-Louis de Lacaille in the middle of the 18th century in memory of the invention of the instrument of revolutionary significance for astronomy. On Lacaille's map it was one of the long lens telescopes of the 17th–18th centuries. At present the constellation lacks Gamma Telescopii which has become G Scorpii in the Scorpion's "sting," as a result of which the present telescope is much shorter than its initial version for which Lacaille borrowed stars from adjacent constellations.

Alpha Telescopii, 3.5 mag, is a bluish-white star of spectral class B3 at a distance of 249 light-years. **Epsilon Telescopii**, 4.5 mag, is a double star with a very faint companion of 13.0 mag at an apparent distance of 21". We are separated from this star by about 410 light-years.

Galaxy **NGC 6845** (opposite below) is the biggest member of a multiple system of galaxies at a distance of some 325 million light-years from our Galaxy. It has a spiral arms over 100,000 light-years long projecting in northeast direction to a small type Sc galaxy.

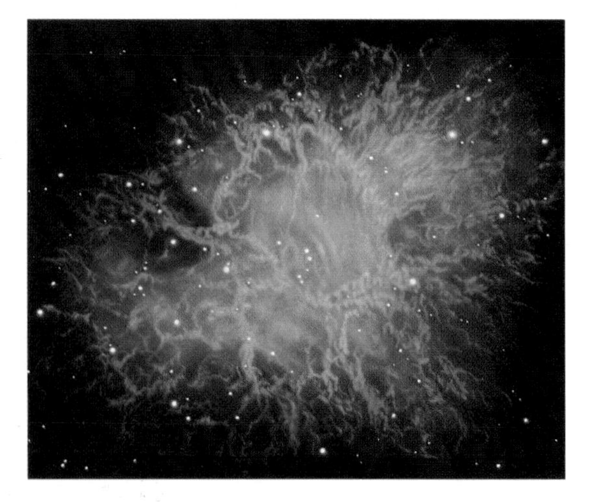

M1 – the Crab Nebula in Taurus, the remnant of a supernova of 1054 (picture to p. 194). The tangle of radial filaments testifies to the enormity of the explosion that generated the Crab. Once these filaments formed the atmosphere of the star. After its explosion, the brightness of the supernova was -5.0 mag, as a result of which the star could be seen even in daylight for 23 days and in the night for 653 days after its appearance.

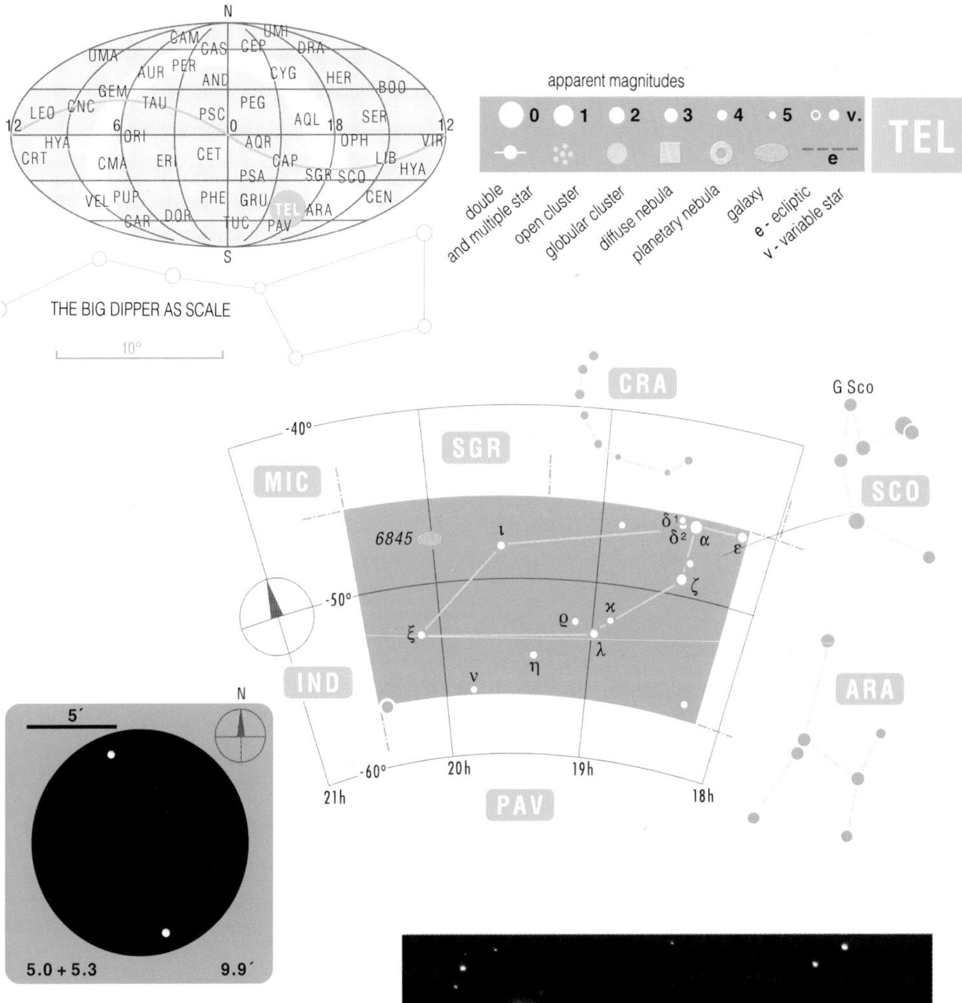

Above: Optical double star **Delta Telescopii** the components of which can be separated with the naked eye.

Right: Group of galaxies in the Telescope. The biggest of them in the picture is the galaxy **NGC 6845**.

TRIANGULUM

Trianguli Tri The Triangle

The small, but distinct Triangle ranks among the classical constellations of ancient times. The Greeks called it Deltotum for its resemblance to their letter delta (△). In his description of constellations, Aratus of Soli noted the similarity of the constellation with the outline of Sicily when he writes: "In the proximity of Andromeda, there is Sicily, which resembles a triangle the shorter side of which is decorated with closely spaced stars."

The three stars marking the corners of the Triangle are situated at considerably different distances. **Alpha Trianguli**, 3.4 mag, has two faint companions of 12.0 and 12.9 mag at a mutual apparent distance of 222" and 83". It shines from a distance of 64 light-years. **Beta Trianguli**, 3.0 mag, is a white giant at a distance of 124 light-years. **Gamma Trianguli**, 4.0 mag, is a star of spectral class A1 at a distance 118 light-years. Let us try to imagine the Triangle in three-dimensional space: it will acquire an entirely different form than it has when projected on the celestial sphere.

The spiral galaxy **M 33 – NGC 598** (below and opposite) of Sc type is the third biggest member of the Local Group of galaxies after our Galaxy and the M 31 galaxy in Andromeda. According to recent studies the distance of M 33 is within the limits of 2.2 to 2.9 million light-years – the second closest spiral to the Milky Way.

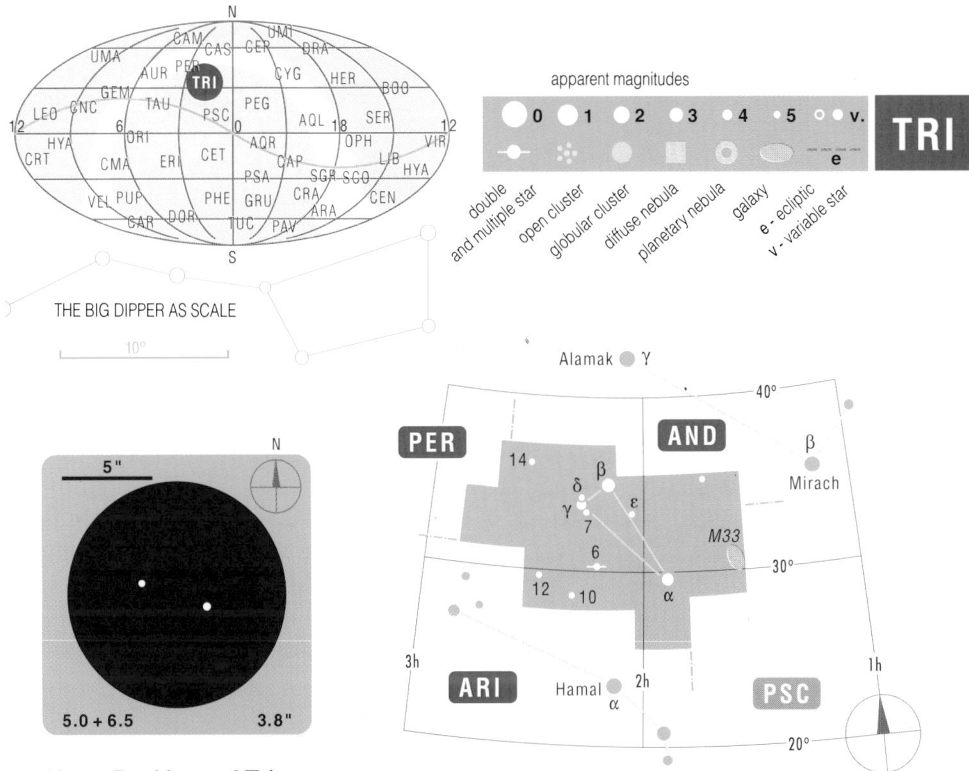

Above: Double star **6 Trianguli** – a physical double with the components of spectral classes G5 and F5.

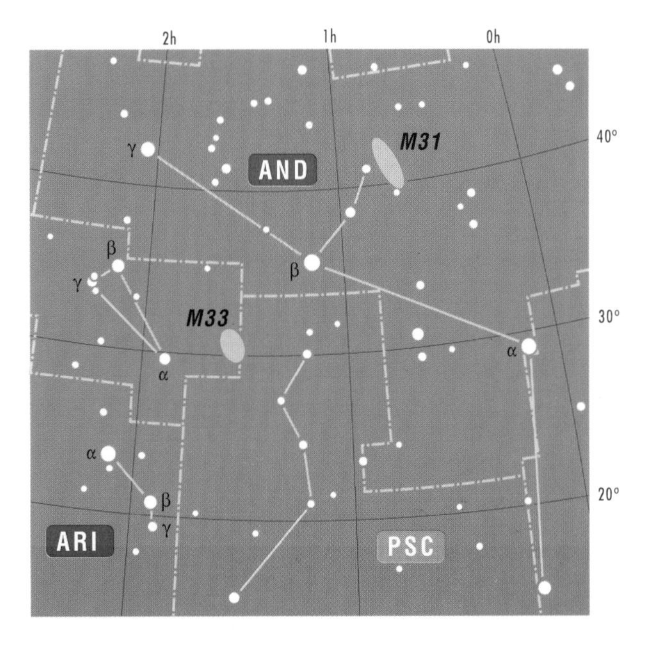

Right: You can find the galaxy M 33 near the line connecting Alpha Trianguli and Beta Andromedae. The apparent diameter of the object is almost 1°. However, its brightness is low and it is difficult to observe. We must use binoculars or a telescope with small magnification and a wide angle of view. The nearby galaxy M 31 is actually situated in space at a distance of some 570,000 light-years from the M 33 galaxy.

TRIANGULUM AUSTRALE

Trianguli Australis TrA The Southern Triangle

East of the "Pointers," Alpha and Beta Centauri, you can find easily a small, but distinct constellation, the Southern Triangle the corners of which are marked with bright stars of second to third magnitude. It is one of the 12 southern constellations with which P. D. Keyser at the end of the 16th century and J. Bayer in 1603 extended the until then recognized set of 48 ancient constellations. In this way the southern sky was provided with a counterpart of the Triangle of the northern sky.

Alpha Trianguli Australis, 1.9 mag, is an orange giant of spectral class K2 and surface temperature of some 4,000 K. The star is situated at a distance of about 415 light-years. **Beta Trianguli Australis**, 2.9 mag, has a faint companion of 14.0 mag at an apparent distance of 155". The principal component is a white star of spectral class F2 at a distance of some 40 light-years. The third vertex of the triangle is occupied by **Gamma Trianguli Australis**, 2.9 mag, a bluish-white star of spectral class A1, shining from the distance of 183 light-years from the Sun.

Amateur Astronomy

Contact with "kindred souls." Within the astronomical community, every amateur astronomer can find highly useful and often essential advice and assistance. Contact with this world and a ready supply of topical information is provided by astronomical periodicals like Astronomy *Magazine and* Sky and Telescope. *There you may also find some useful addresses of the astronomy clubs, associations, observatories and planetariums, INTERNET pages, etc.*

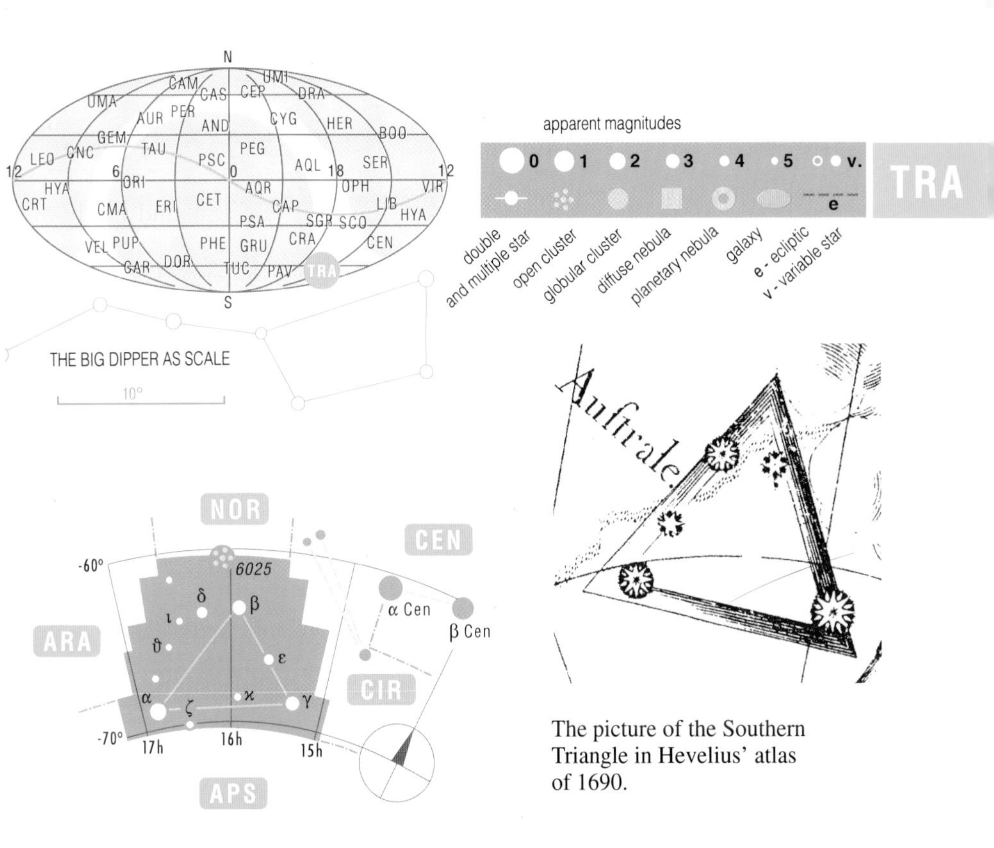

The picture of the Southern Triangle in Hevelius' atlas of 1690.

Right: Open star cluster **NGC 6025**, of total brightness 5.5 mag. Its brightest stars are of 7.0 mag. Within a circle of 15' there are about 140 stars. The distance of the cluster is estimated at 2,700 light-years.

The exotic bird with a powerful beak rose to the sky on the proposal of P. D. Keyser and thanks to J. Bayer who included the constellation of Toucan in his *Uranometria* in 1603. This not very conspicuous constellation can be found with the assistance of the Small Magellanic Cloud (SMC) situated near the southern border of the Toucan at an apparent distance of less than 20° from the south celestial pole.

Beta Tucanae is a sextuple system. A small telescope reveals a pretty double star with the components of Beta 1 and Beta 2 of 4.4 and 4.5 mag at a mutual distance of 27". At an apparent distance of 9.3' to the southeast there is the third component, Beta 3, of 5.1 mag. All three components are close binaries.

The **Small Magellanic Cloud** (SMC) consists of two irregular dwarf galaxies visible in the same direction from the Earth. Their mutual distance is about 20,000 light-years. They are connected gravitationally with our Galaxy. The distance of the two SMC galaxies from the Earth is about 200,000 light-years. Two bright globular star clusters, **NGC 104** and **NGC 362**, can be seen in the direction of SMC. However, this is only apparent, as actually they are much nearer and they belong to our Galaxy. The distance of the NGC 362 star cluster from us is about 32,000 light-years.

Globular cluster **47 Tucanae – NGC 104** is the second brightest cluster of its type (after Omega Centauri). It can be observed with the naked eye as a hazy starlet of 5.0 mag. The brightest stars in the cluster are of 11.5 mag. Therefore you need a bigger telescope for their visual separation. The distance of the system is estimated within the limits of 13,000 and 20,000 light-years.

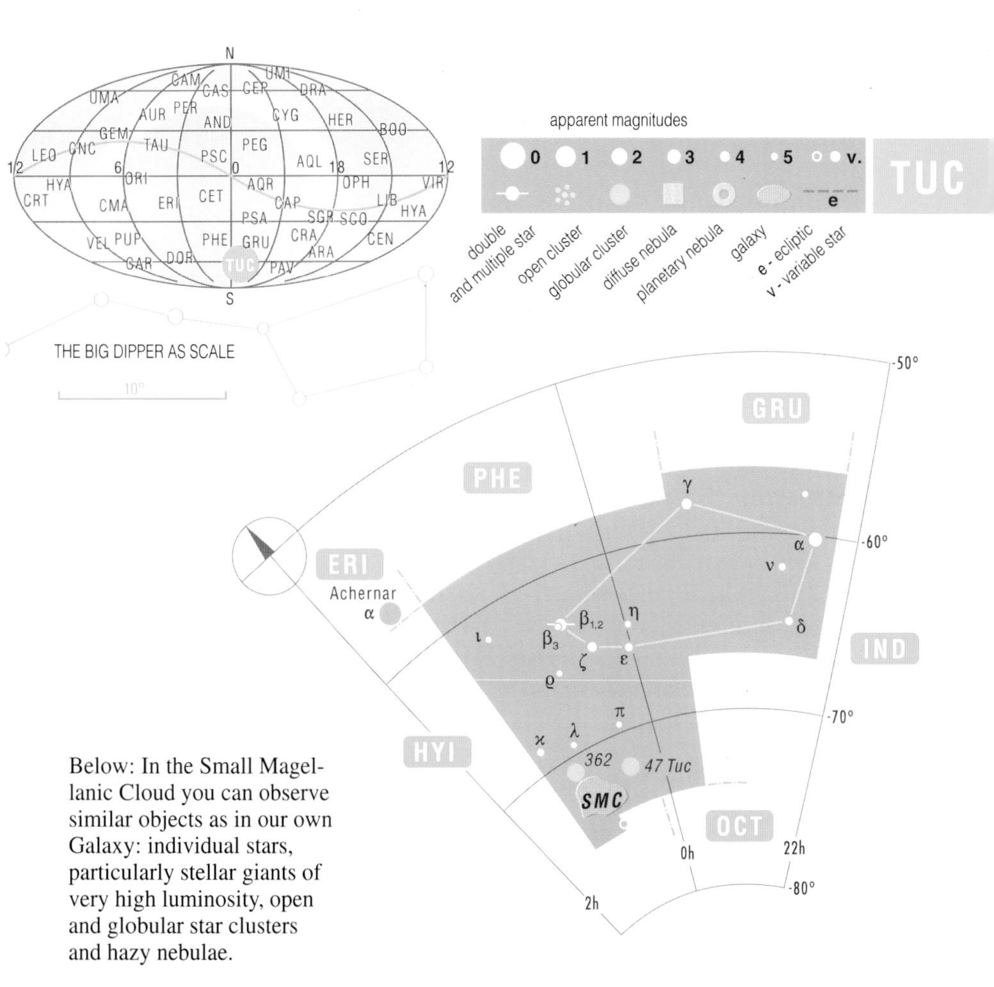

Below: In the Small Magellanic Cloud you can observe similar objects as in our own Galaxy: individual stars, particularly stellar giants of very high luminosity, open and globular star clusters and hazy nebulae.

URSA MAJOR

Ursae Majoris UMa The Great Bear

One of the oldest and best-known constellations accompanied by a great number of legends as well as different names in various nations. In area it is the third largest constellation of the sky. The asterism of the seven brightest stars of the Great Bear is known as the Big Dipper, the Plough, and the Wain. (An asterism is an easily recognized group of stars within a constellation.)

The pattern of the Big Dipper is not an accidental group of seven stars: apart from Dubhe and Benetnash, the remaining five stars, together with twelve further stars situated in different parts of the sky, form the **Ursa Major moving cluster**, the nearest of all star clusters: its center is situated some 80 light-years from us. Its diameter is about 30 light-years. The system is moving at 14 km per second in the direction of the Archer (Sagittarius). The Hyades in the Bull (Taurus) are a similar formation (p. 195).

Mizar and Alcor are the best-known multiple star. With the naked eye you can see two stars: **Zeta UMa – Mizar**, 2.4 mag, and the fainter **80 UMa – Alcor**, 4.0 mag, 11.8′ to the east. The distance of Mizar is 78 light-years; that of Alcor is 81 light-years.

The pair of galaxies, **M 81 – NGC 3031** (above) and **M 82 – NGC 3034** (below) form the nucleus of a 12-member group of galaxies some 11 million light-years away from us. Some 200 million years ago, M 81 passed in the proximity of M 82. The influence of their mutual gravitation resulted in the unusual structure and activity of M 82. Both galaxies are mutually connected by a gaseous "bridge." Their apparent distance in the sky is 38′. They are visible even with small telescopes.

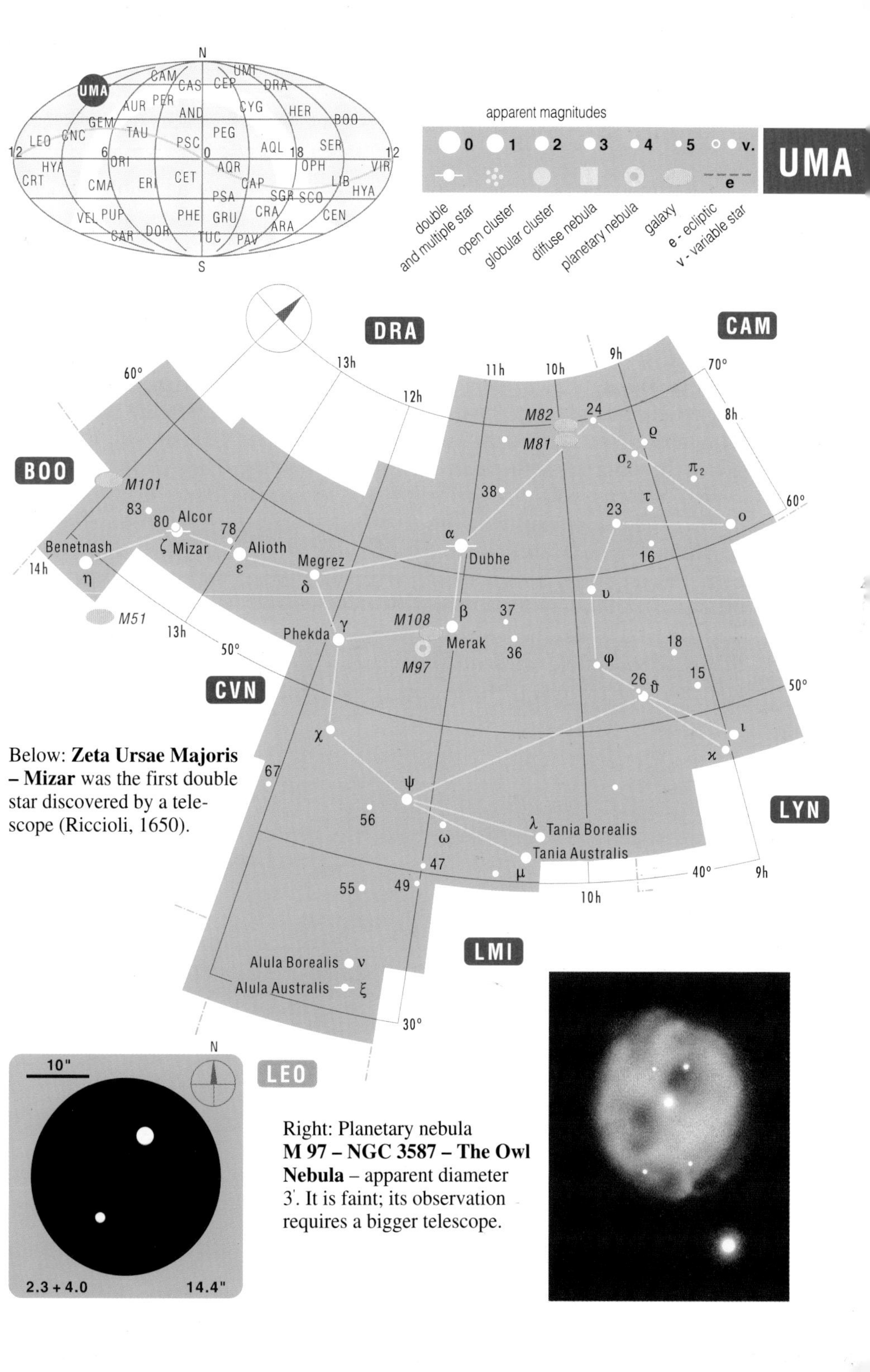

Below: **Zeta Ursae Majoris – Mizar** was the first double star discovered by a telescope (Riccioli, 1650).

Right: Planetary nebula **M 97 – NGC 3587 – The Owl Nebula** – apparent diameter 3'. It is faint; its observation requires a bigger telescope.

URSA MINOR

Ursae Minoris UMi The Little Bear

The constellation contains one of the best-known stars – the North Star or Polaris – the best orientation point when looking for the celestial north pole. With six fainter stars the North Star forms the asterism of the Little Dipper, much less conspicuous than the majestic Big Dipper (the Great Wain). The same asterism decorates the picture of the Little Bear with the North Star at the end of his unbelievably long tail. According to ancient Greek mythology, the long tails of both Bears originated when Zeus pulled them up among the constellations in the sky. Why should he do so? Because the philandering Zeus had a son named Arcas with the beautiful nymph Callisto. Hera, Zeus' jealous wife, changed Callisto into a bear, which the ignorant Arcas nearly killed when hunting. The killing was prevented by Zeus, who changed Arcas into another bear and immortalized mother and son as constellations.

Alpha Ursae Minoris – Polaris (the North Star), 2.1 mag, was known earlier as a pulsating variable star – cepheid, changing its brightness by 0.12 mag within four days. However, the pulsations diminished continuously until they attained 0.02 mag in the middle of the 1990s. The North Star is also a double star with a companion of 9.0 mag at an apparent distance of 18.4". Alpha UMi is a supergiant some 430 light-years away.

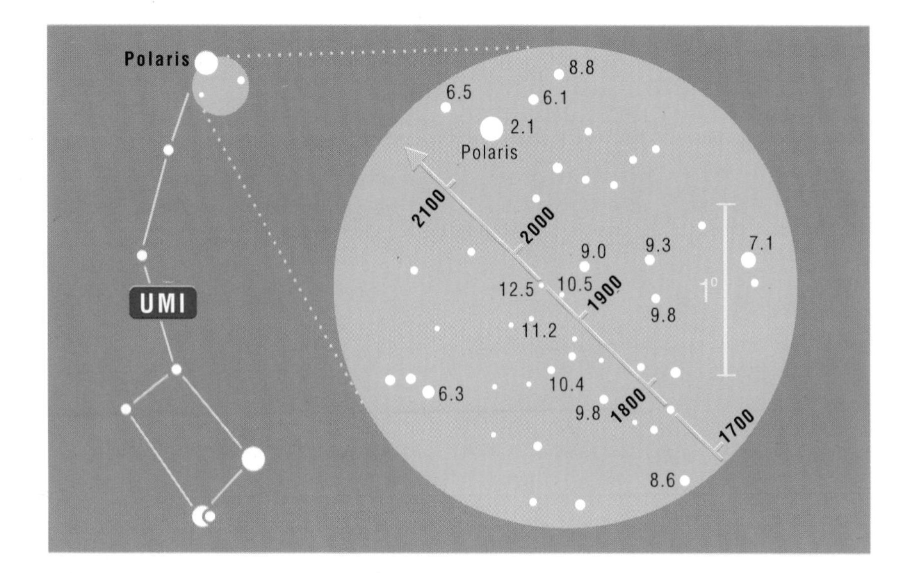

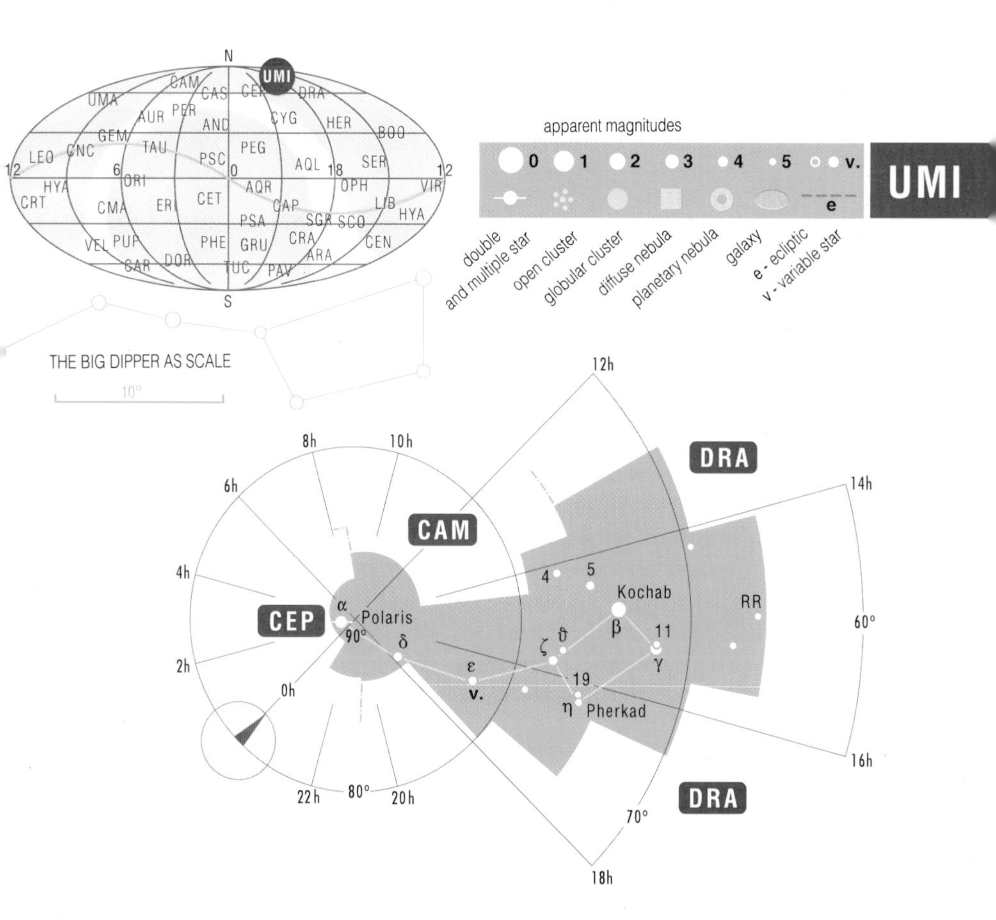

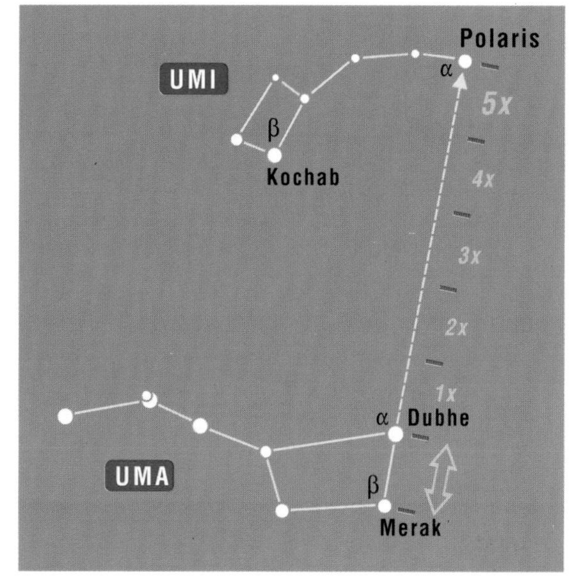

Left: Position of the celestial north pole among the stars varies due to the precession of the Earth's axis (see p. 58). In 2102, the North Star will be nearest the pole (27' 37"). The faint stars with indicated magnitudes can be used for the estimate of the faintest magnitude visible with a given telescope.

Right: We can find the North Star on the line connecting the stars Merak and Dubhe at five times their mutual distance, as shown in the picture.

VELA

Velorum Vel The Sails

The constellation is one of the remainders of the Ship Argo divided by Lacaille in the middle of the 18th century into smaller constellations (see also Carina, Puppis, Pyxis).

Gamma Velorum (see the opposite page) is a multiple system in which a bright double star with components A and B, separable even with binoculars, is outstanding. Component A is a close binary comprising the brightest-known Wolf-Rayet Star, an extraordinarily luminous (15,000 suns) and hot (30,000 K) giant with a fast-expanding gaseous envelope.

NGC 3132 is one of the brightest planetary nebulae at the boundary between Vela and Antlia (see p. 48).

The **remnant of supernova SNR Vela** (picture below) looks like a fine fabric of radiant fibers covering an area of some 6° in diameter. The supernova exploded some 12,000 years ago. The star brightened 100 million times and became the brightest celestial object, second only to the Moon. In the proximity of the center of the expanding nebula, there is a so-called pulsar, a speedily rotating (11 revolutions per second) neutron star of only 10 to 15 km in diameter – the remnant of the initial star.

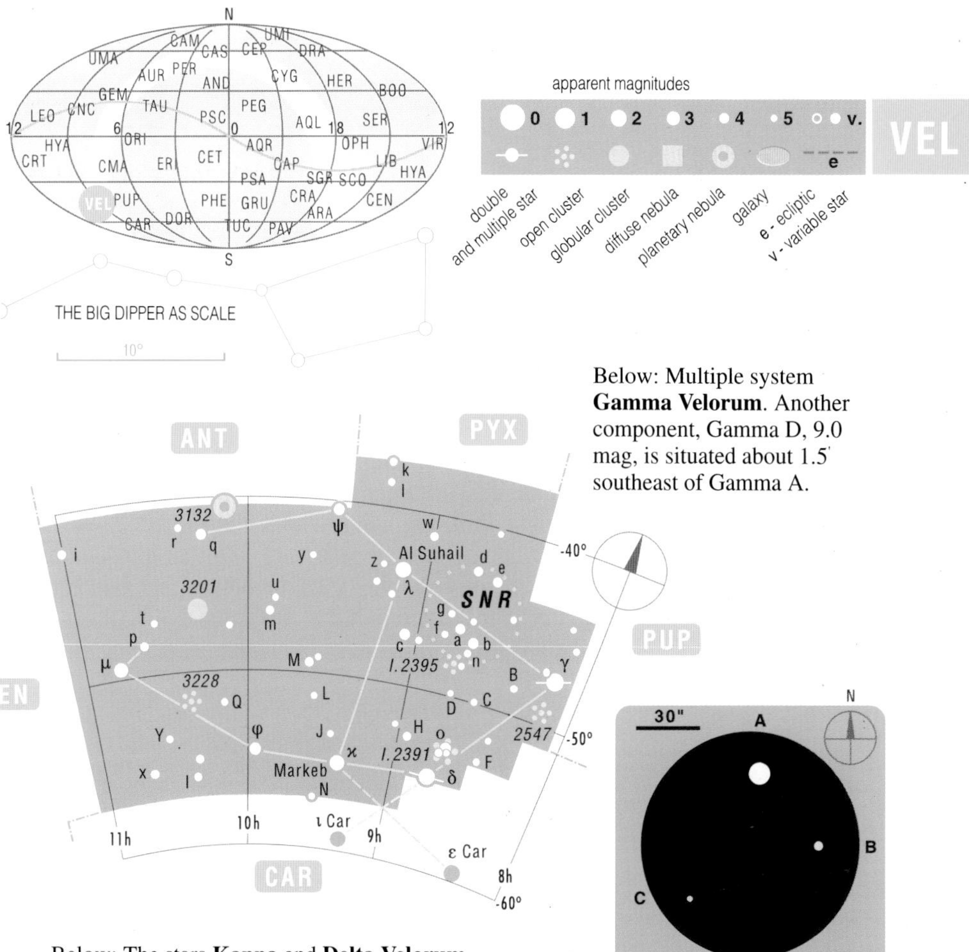

Below: Multiple system **Gamma Velorum**. Another component, Gamma D, 9.0 mag, is situated about 1.5′ southeast of Gamma A.

Below: The stars **Kappa** and **Delta Velorum**, together with Iota and Epsilon Carinae, form the conspicuous “false cross.” The genuine Southern Cross is indicated by the Pointers – the stars Alpha and Beta Centauri.

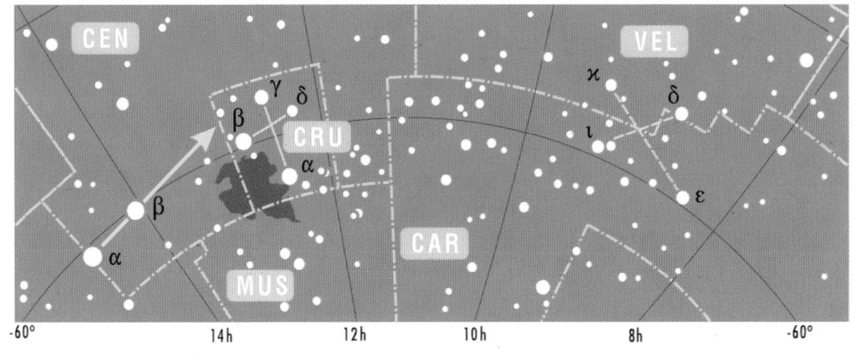

VIRGO

Virginis Vir The Virgin

In mythology this celestial lady is connected with the goddesses of justice or crops. The justice is symbolized by the adjacent Scales (Libra). As a symbol of crops, the Virgin holds corn ears in the area of the bright star Spica (spica = corn ear in Latin). In respect of area, the Virgin is the second largest constellation.

Alpha Virginis – Spica, 0.9 mag, with variations of about 0.05 mag. The luminosity of the star is 1,600 suns, its distance is 262 light-years. **Gamma Virginis – Porrima**, total brightness 2.9 mag, is a very pretty double (see below) 39 light-years away.

The extensive **cluster of galaxies in Virgo** extends as far as beyond the boundaries of the adjacent Berenice's Hair (Coma Berenices). Its central part, numbering some 3,000 members, is shown by the dotted circle on the map on the right. It is the nearest of big clusters of galaxies. As a result, several dozens of brighter galaxies are within reach of amateur telescopes. Although the cluster is relatively near, its distance estimations by various methods vary from 45 to 80 million light-years.

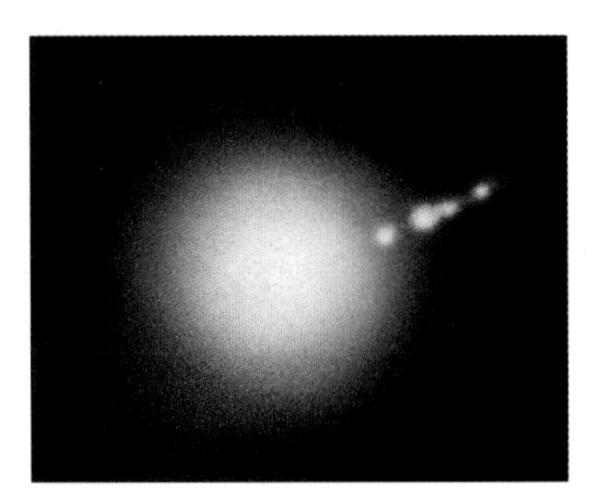

Above: **M 87 – NGC 4486**, one of the biggest members of the cluster of galaxies in Virgo, a strong source of radio and X-ray radiation, known as **Virgo A**. The picture shows a jet of matter extending from the nucleus of the galaxy. The orbiting motion of the stars of the galaxy implies the existence of a black hole in its nucleus.

Below: The period of the binary **Gamma Virginis** is 171 years. The components are closing in at present.

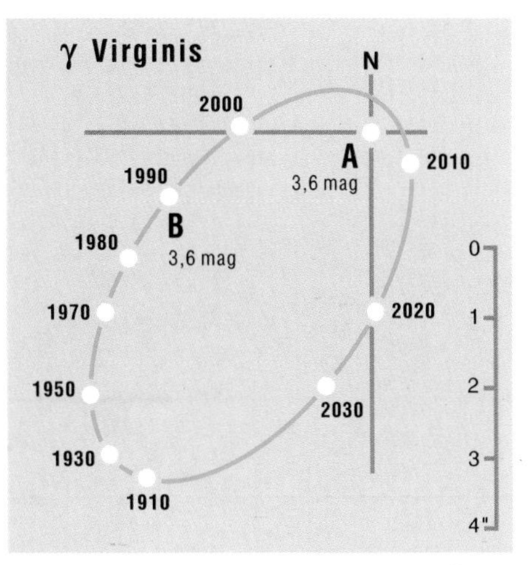

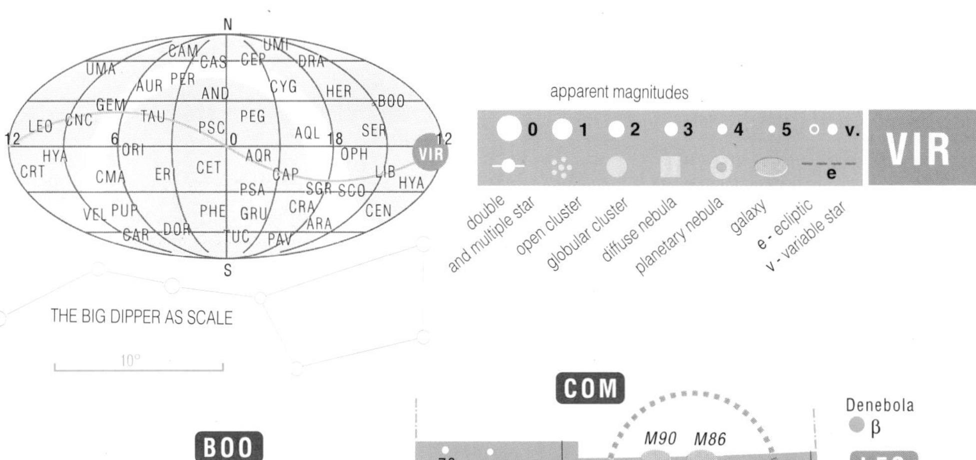

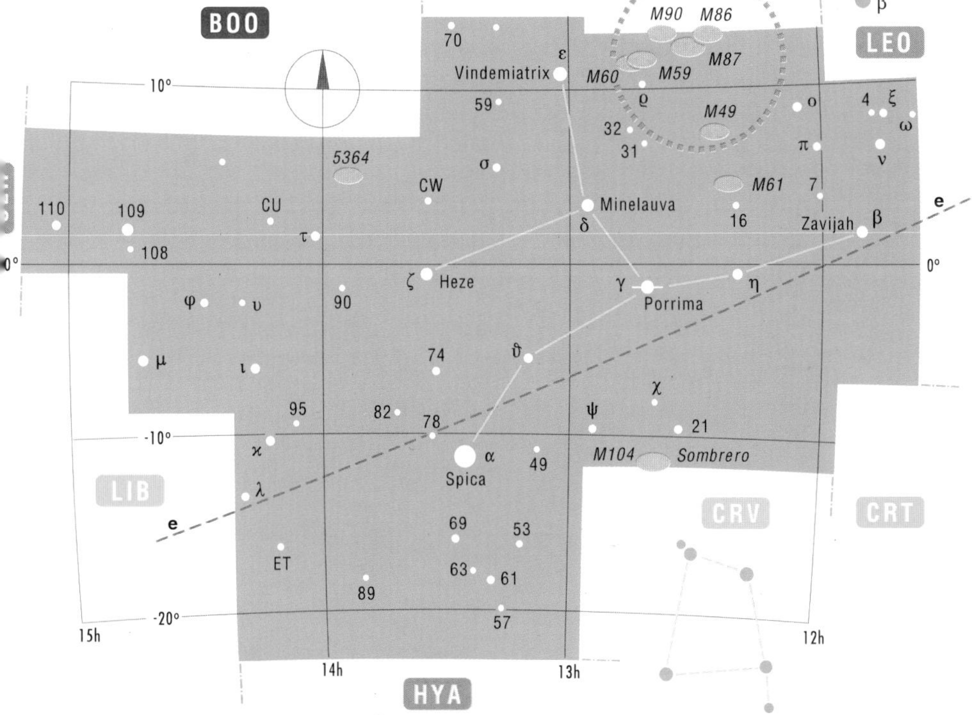

Left: The galaxy **M 104 – NGC 4594 – The Sombrero Galaxy** – a perfect example of the presence of a band of dark interstellar matter along the equatorial plane of the galaxy.

VOLANS

Volantis Vol The Flying Fish

The constellation was introduced by Johann Bayer in his atlas *Uranometria* in 1603 under the name of Piscis Volans. It is one of the creations of the Dutch navigator P. D. Keyser. It is probably not accidental that the Flying Fish are accompanying a ship, the ship Argo, which formed an integral constellation still in Keyser's time, before being "carved" into smaller constellations by Lacaille. The inconspicuous Flying Fish can be found in the neighborhood of the Large Magellanic Cloud, in the direction of the "false cross" (see Vela).

Alpha Volantis, 4.0 mag, is a star of spectral class A at a distance of 124 light-years. **Beta Volantis** is an orange giant of spectral class K2, 108 light-years away. The yellow supergiant **Delta Volantis**, 4.0 mag, shines some 660 light-years away. The components of the double star **Epsilon Volantis**, of 4.3 and 7.4 mag, situated at a mutual apparent distance of 6", can be separated with a small telescope.

The picture below shows the galaxy **NGC 2442** of the SBb type, with two distinct spiral arms and a small nucleus. The galaxy contains large quantities of dust, creating clouds which are particularly visible in the broad band in the northern arm.

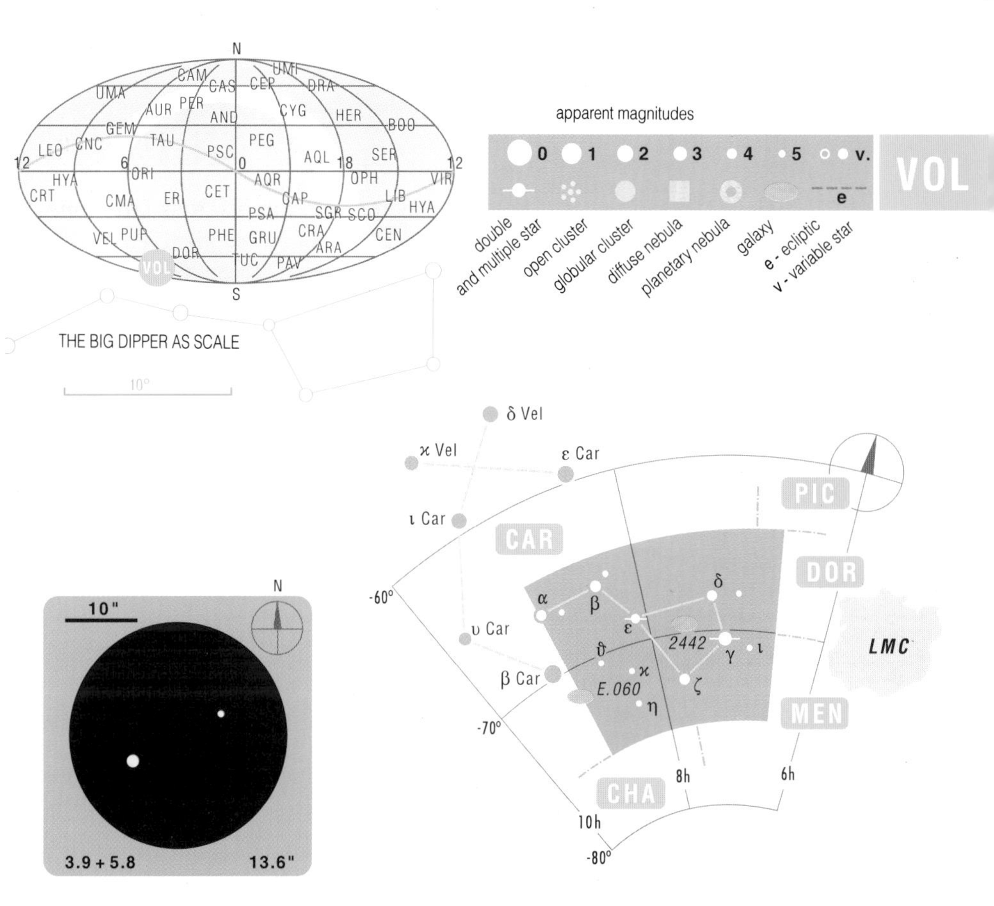

Above: The double star **Gamma Volantis** is a pretty object, which can be observed with a small telescope. Its brighter component is yellowish.

Right: The small group of galaxies **ESO 060**, discovered during a sky survey of the European Southern Observatory (ESO) in Chile. The big galaxy in its center shows the traces of gravitational influence, most probably after a meeting with an adjacent galaxy.

VULPECULA

Vulpeculae Vul The Fox

One of the not very pretty constellations introduced by Johannes Hevelius from Gdansk (Danzig) in 1690, initially with the name of Vulpecula cum Anser (fox with a goose). It can be found in the neighborhood of the Swan (Cygnus). In ancient atlases we can really find a goose, dangling haplessly by the neck from the toothy muzzle of the running fox. In modern constellation boundaries, only the fox has remained. There is no myth explaining what happened to the goose and it is left to the reader's imagination to fill the story of this alphabetically last constellation.

The group of stars **C 399** (Collinder 399), known as the **Coathanger**, is a pretty object for the binoculars. The picture opposite below shows that six stars in a line form the bar (1.5° long) and another four stars the hook of the coathanger. The brightness of these ten stars varies between 5.0 and 7.0 mag.

M 27 – NGC 6853, known as the **Dumbbell Nebula** (see below), ranks among the biggest and brightest planetary nebulae. It can be seen even with binoculars. Its apparent dimensions are $5' \times 8'$.

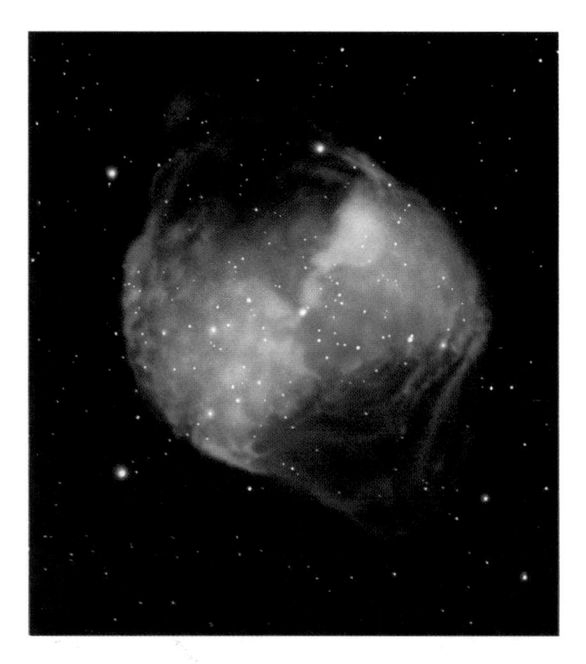

Planetary nebula **M 27 – the Dumbbell Nebula** – expands at the rate of some 30 km/s. Its distance is estimated by various authors within the limits of 300 and 900 light-years. The central star of 14.0 mag has a temperature of about 85,000 K.

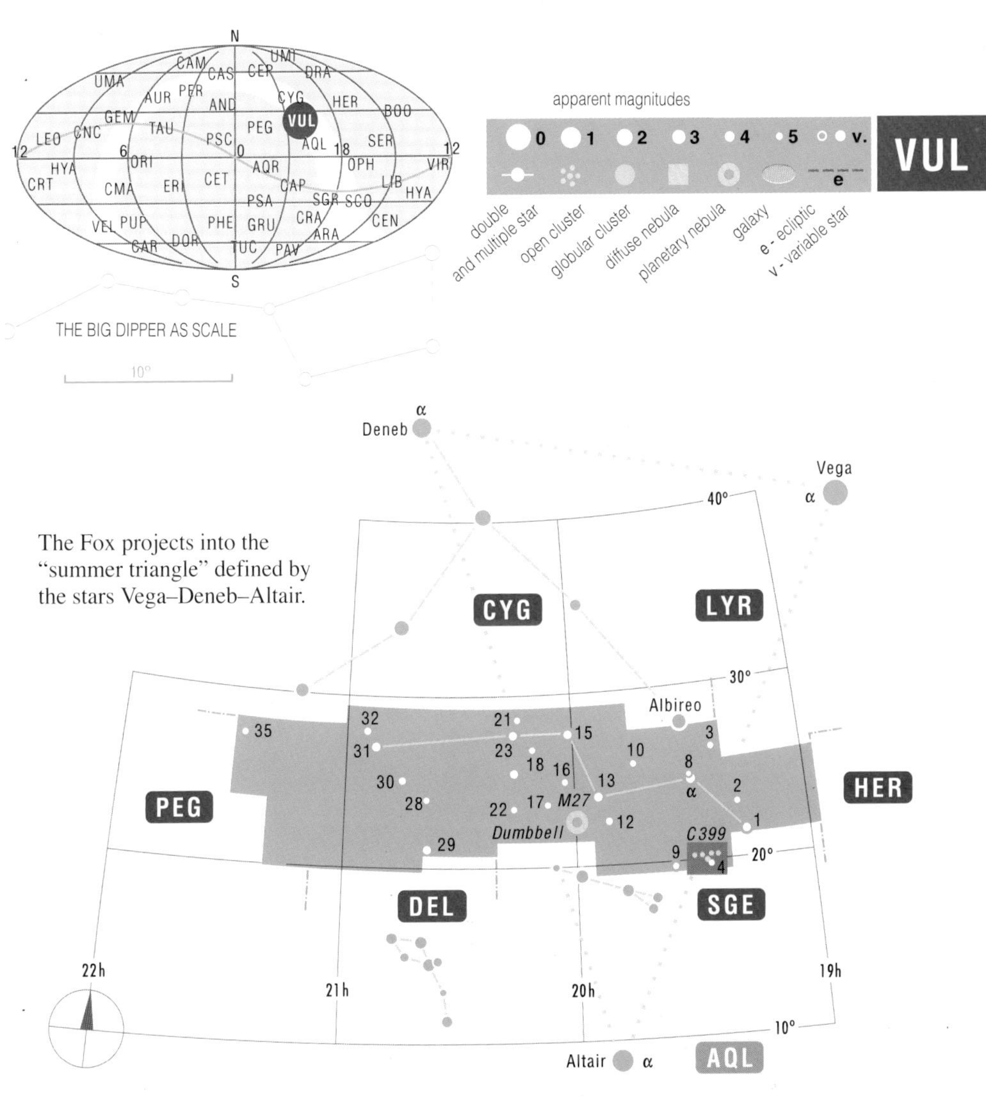

The Fox projects into the "summer triangle" defined by the stars Vega–Deneb–Altair.

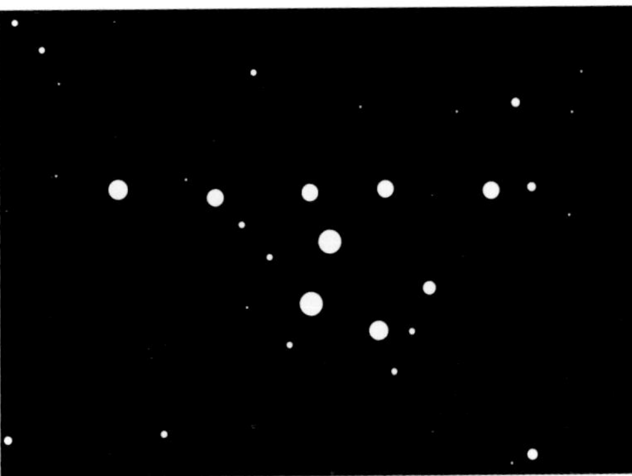

Right: The star group called **Coathanger** was formerly classified as open cluster C 399 in the Collinder's Catalogue. Accurate measurements of stellar distances made from HIPPARCOS satellite revealed C 399 as random group of stars at very different distances from the Earth.

INDEX OF CONSTELLATIONS

Latin name	genitive form	abbreviation	common name	page
Andromeda	*Andromedae*	And	Andromeda	46
Antlia	*Antliae*	Ant	Air Pump	48
Apus	*Apodis*	Aps	Bird of Paradise	50
Aquarius	*Aquarii*	Aqr	Water Carrier	52
Aquila	*Aquilae*	Aql	Eagle	54
Ara	*Arae*	Ara	Altar	56
Aries	*Arietis*	Ari	Ram	58
Auriga	*Aurigae*	Aur	Charioteer	60
Bootes	*Bootis*	Boo	Herdsman	62
Caelum	*Caeli*	Cae	Chisel	64
Camelopardalis	*Camelopardalis*	Cam	Giraffe	66
Cancer	*Cancri*	Cnc	Crab	68
Canes Venatici	*Canum Venaticorum*	CVn	Hunting Dogs	70
Canis Major	*Canis Majoris*	CMa	Great Dog	72
Canis Minor	*Canis Minoris*	CMi	Little Dog	74
Capricornus	*Capricorni*	Cap	Goat	76
Carina	*Carinae*	Car	Keel	78
Cassiopeia	*Cassiopeiae*	Cas	Cassiopeia	80
Centaurus	*Centauri*	Cen	Centaur	82
Cepheus	*Cephei*	Cep	Cepheus	84
Cetus	*Ceti*	Cet	Whale	86
Chamaeleon	*Chamaeleontis*	Cha	Chameleon	88
Circinus	*Circini*	Cir	Compasses	90
Columba	*Columbae*	Col	Dove	92
Coma Berenices	*Comae Berenices*	Com	Berenice's Hair	94
Corona Australis	*Coronae Australis*	CrA	Southern Crown	96
Corona Borealis	*Coronae Borealis*	CrB	Northern Crown	98
Corvus	*Corvi*	Crv	Crow	100
Crater	*Crateris*	Crt	Cup	100
Crux	*Crucis*	Cru	Southern Cross	102
Cygnus	*Cygni*	Cyg	Swan	104
Delphinus	*Delphini*	Del	Dolphin	106
Dorado	*Doradus*	Dor	Swordfish	108
Draco	*Draconis*	Dra	Dragon	110
Equuleus	*Equulei*	Equ	Colt *or* Foal	112
Eridanus	*Eridani*	Eri	River Eridanus	114
Fornax	*Fornacis*	For	Furnace	116
Gemini	*Geminorum*	Gem	Twins	118
Grus	*Gruis*	Gru	Crane	120
Hercules	*Herculis*	Her	Hercules	122
Horologium	*Horologii*	Hor	Pendulum Clock	124
Hydra	*Hydrae*	Hya	Water Snake	126
Hydrus	*Hydri*	Hyi	Lesser Water Snake	128
Indus	*Indi*	Ind	Indian	130
Lacerta	*Lacertae*	Lac	Lizard	132
Leo	*Leonis*	Leo	Lion	134
Leo Minor	*Leonis Minoris*	LMi	Lesser Lion	134

Latin name	genitive form	abbreviation	common name	page
Lepus	*Leporis*	Lep	Hare	136
Libra	*Librae*	Lib	Scales *or* Balance	138
Lupus	*Lupi*	Lup	Wolf	140
Lynx	*Lyncis*	Lyn	Lynx	142
Lyra	*Lyrae*	Lyr	Lyre	144
Mensa	*Mensae*	Men	Table Mountain	146
Microscopium	*Microscopii*	Mic	Microscope	148
Monoceros	*Monocerotis*	Mon	Unicorn	150
Musca	*Muscae*	Mus	Fly	152
Norma	*Normae*	Nor	Rule	154
Octans	*Octantis*	Oct	Octant	156
Ophiuchus	*Ophiuchi*	Oph	Serpent Bearer	158
Orion	*Orionis*	Ori	Orion *or* Hunter	160
Pavo	*Pavonis*	Pav	Peacock	162
Pegasus	*Pegasi*	Peg	Pegasus	164
Perseus	*Persei*	Per	Perseus	166
Phoenix	*Phoenicis*	Phe	Phoenix	168
Pictor	*Pictoris*	Pic	Painter	170
Pisces	*Piscium*	Psc	Fishes	172
Piscis Austrinus	*Piscis Austrini*	PsA	Southern Fish	174
Puppis	*Puppis*	Pup	Stern	176
Pyxis	*Pyxidis*	Pyx	Mariner's Compass	176
Reticulum	*Reticuli*	Ret	Reticle	178
Sagitta	*Sagittae*	Sge	Arrow	180
Sagittarius	*Sagittarii*	Sgr	Archer	182
Scorpius	*Scorpii*	Sco	Scorpion	184
Sculptor	*Sculptoris*	Scl	Sculptor	186
Scutum	*Scuti*	Sct	Shield	188
Serpens	*Serpentis*	Ser	Serpent	190
Sextans	*Sextantis*	Sex	Sextant	192
Taurus	*Tauri*	Tau	Bull	194
Telescopium	*Telescopii*	Tel	Telescope	196
Triangulum	*Trianguli*	Tri	Triangle	198
Triangulum Australe	*Trianguli Australis*	TrA	Southern Triangle	200
Tucana	*Tucanae*	Tuc	Toucan	202
Ursa Major	*Ursae Majoris*	UMa	Great Bear	204
Ursa Minor	*Ursae Minoris*	UMi	Little Bear	206
Vela	*Velorum*	Vel	Sails	208
Virgo	*Virginis*	Vir	Virgin	210
Volans	*Volantis*	Vol	Flying Fish	212
Vulpecula	*Vulpeculae*	Vul	Fox	214

GLOSSARY OF TERMS

Absolute Temperature
The lowest temperature theoretically possible. It is given in Kelvin's (K). Absolute zero = 0 K = -273.16 °C. Freezing point = 273.16 K = 0 °C.
Absorption (of radiation)
The absorption of radiation by the atoms or molecules of gases. In broader meaning, any weakening of radiation along the path of a radiation beam from the source to the observer.
Angular Measure
A whole circle is divided into 360° (degrees), which is subdivided into 60' (minutes) each. Every minute is subdivided in turn into 60" (seconds). This scale is used for the dimensions of objects and their mutual distances on a celestial sphere. A good example and scale is the diameter of the full Moon, which is approximately 0.5° = 30'.
Astronomical Catalogues
Lists of celestial objects with their coordinates and principal characteristics. Particularly popular are the catalogues of non-stellar objects used most frequently for the denomination of star clusters, nebulae and galaxies:
M – Messier's Catalogue
NGC – New General Catalogue
For instance: M 31 – NGC 224, The Great Andromeda Galaxy.
Astronomical Unit (AU)
Average distance of the Earth from the Sun, 149.6 million kilometers. It is used as a unit of distance in the solar system. One *light-year* = 63,300 AU.
Black Hole
The remains of a collapsed star the mass of which exceeds that of two suns. The black hole is black because is does not emit any radiation. However, it exercises powerful gravitational effects on its environs.
Cluster of Galaxies
A large system comprising several hundreds or even thousands of galaxies. Several thousands of galaxy clusters are known, such as the cluster of galaxies in Virgo (the Virgin).
Constellation
Accurately defined part of the sky. In the narrower meaning of the term, a group of (bright) stars forming a characteristic pattern. The whole sky is divided into 88 constellations, the boundaries of which were defined by the International Astronomical Union (IAU).
Deep-Sky Objects
Collective denomination of the objects in the stellar universe, star clusters, nebulae, galaxies, etc., in current use in astronomical literature.
Double Star (Double, Binary)
Optical Double – two stars that appear very close to each other in the sky, but do not form any physical system.
Physical Double – a system of two stars mutually bound by gravitational forces and orbiting around a common center of gravity.
Spectroscopic Binary – two very close components which cannot be separated by telescope, but are betrayed by the doubling of spectral lines.
Colored Double – differences of temperature and brightness of components

together with the physiology of colored vision create contrasting colors of the two components (see e.g., Beta Cygni– Albireo).
Visual Double – a double star the components of which can be separated by telescope. The components of double stars are usually denominated A (brighter) and B (fainter), in case of multiple stars also C, D, etc.
Eclipsing Binary – a double star the components of which cannot be separated by telescope; however, their mutual regular eclipsing is indicated by periodic variation of brightness (see, e.g., Beta Persei – Algol).

Galaxy

Our Galaxy – stellar system of which the Sun is a member. The surrounding parts of our Galaxy project on the sky in the form of the Milky Way.
External Galaxies – stellar systems analogous with our Galaxy.

Giant, Supergiant

Star of very high luminosity and much (ten to thousand times) greater diameter than the Sun.

Greek Alphabet

Small letters of Greek alphabet are used as denomination of bright stars in individual constellations. This practice was introduced by Johann Bayer in his stellar atlas *Uranometria* in 1603.

α	alpha	ι	iota	ρ	rho
β	beta	κ	kappa	σ	sigma
γ	gamma	λ	lambda	τ	tau
δ	delta	μ	mu	υ	upsilon
ε	epsilon	ν	nu	φ	phi
ζ	zeta	ξ	xi	χ	chi
η	eta	ο	omicron	ψ	psi
ϑ	theta	π	pi	ω	omega

Interstellar Gas

Gaseous component of interstellar matter consisting of electrons, ions, atoms and molecules. Some 99% of interstellar gas consists of hydrogen and helium. The basic forms of interstellar gas clouds are H-I and H-II regions (H = hydrogen).
H-I regions contain cold, non-luminous clouds of neutral hydrogen. The clouds are not visible, but emit radio radiation at a wavelength of 21 cm; therefore, they can be observed with radiotelescopes.
H-II regions contain clouds of ionized hydrogen in the vicinity of very hot stars of spectral types O and B. If the gas is of sufficient density, the regions appear as luminous nebulae.

Light Curve

Diagram showing the change of brightness of a celestial object (e.g., a variable star).

Light-year

Unit of distance used particularly in popular astronomical literature. One light-year is the distance traversed by light in a year. The velocity of light being approximately 300,000 km/s, it holds that:

1 light-year = 9,460,000,000,000 km =
= $9.46 . 10^{12}$ km = 9.5 trillion km

1 light-year = 63,300 AU = 0.307 pc

Luminosity

Total energy radiated from the surface of a star (stellar system) per time unit.

For clarification it is often defined in terms of the luminosity of the Sun. For instance, the star has the L of 50 suns: its luminosity is 50 times as high as that of the Sun.

Magnitude

From the Latin magnitude; symbol: mag.

The scale of brightness of stars. The faintest stars visible to the naked eye are of about 6 mag, or have the 6th magnitude. Very bright stars are of 1.0 mag (Spica), still brighter stars of 0.0 mag (Vega), and the brightest stars are of negative magnitudes (e.g., Sirius -1.4 mag). The star of a certain magnitude is 2.512 brighter as the star of one magnitude fainter. For instance, a star of 1.0 mag is 100 times brighter than a star of 6.0 mag ($2.512^5 = 100$). Unless stated otherwise, the term "magnitude" always means apparent magnitude, the brightness of the star as it appears to us in the sky.

Nebula (galactic nebula)

Gas or gas and dust formation in interstellar space within a galaxy.

Dark Nebula – appears dark on the background of luminous nebulae or star fields.

Diffuse Nebula -- nebula of irregular shape. It can be an emission or reflection nebula, an H-II region or a combination of these types.

Emission Nebula – the gas is made luminous by the radiation of nearby hot stars.

Planetary Nebula – large luminous envelope surrounding a hot star. In the telescope it sometimes appears as a fuzzy planet – hence the name.

Reflection Nebula – cloud of dust made visible by scattered light of nearby stars.

Neutron Star

Final evolution stage of a star of initial mass equal 1.4–2.5 times the mass of the Sun. Its gravitational collapse reduces it to a sphere of 10–20 km in diameter consisting mostly of free neutrons. The density of a neutron star is hundred billion times (10^{14}) higher than water. Some rotating neutron stars appear as *pulsars*, as fast pulsating sources of radiation.

Nova

Star whose brightness suddenly increases by 7.0–16.0 mag and subsequently decreases to its initial brightness. The phenomenon is seen in white dwarfs forming close binaries with "ordinary" main sequence stars. The flare-up of a nova is due to the explosion of gaseous material flowing from the other star to the white dwarf.

Parallax

The apparent shift in position of an object when viewed from two different positions.

Parsec (pc)

Basic unit of distance in astronomy. It is a distance from which one astronomical unit (AU, the radius of the Earth's orbit) appears under the angle of 1 second of arc.

1 pc = 206,265 astronomical units (AU) = 3.26 light-years
1,000 pc = 1 kpc = 1 kiloparsec
1,000,000 pc = 1 Mpc = 1 megaparsec

Proper Motion (of a star)

Shift of position of a star in the sky due to its actual motion in space. Barnard's Star in the Serpent Bearer (Ophiuchus) has the highest proper motion of 10.34" per year.

Star

Celestial body shining with its own light, thanks particularly to thermonuclear reactions taking part in its interior. A star consists mostly of plasma, a gaseous mixture of freely floating electrons and ionized atoms.

Star Cluster

System of stars of common origin mutually bound by gravitational forces.

Open (star) cluster (or *galactic cluster*) is of irregular shape and contains a smaller number of stars, mostly young (hundreds of million years old).

Globular (star) cluster is of regular spherical shape with star concentration in its center; it contains millions of stars and its age exceeds sometimes 10 trillion years.

Supernova

An explosion of a star during which its luminosity increases as many as ten billion times. The explosion takes place during the gravitational collapse of a very massive star which becomes a neutron star. The cast-off material forms a *supernova remnant* in the form of an expanding nebula (the Crab Nebula in Taurus).

Telescope

Optical instrument for astronomical observations. The objective forms the image of the object which we observe through a magnifying lens – the eyepiece. There are two principal telescope types:

Refractor (lens telescope) – the objective consists of a converging (positive) lens or a system of such lenses,

Reflector (mirror telescope) – the objective consists of a concave – spherical or parabolic – mirror.

Variable Star

Star of variable brightness. Principal types of variable stars are:

Pulsating variables, periodically increasing and decreasing their volume.

Eruptive variables, changing their brightness suddenly, often considerably (novae, supernovae) and generally quite irregularly.

Eclipsing variables, or *eclipsing binaries*, a special group of variables the brightness change of which is due to periodic eclipsing of one components by the other.

White Dwarf

Collapsed star in the final stage of evolution. It consists of degenerate gas, i.e., crushed atoms the nuclei of which approached one another, thus increasing enormously the density of the star. White dwarfs have the dimensions of planets, but the mass of the Sun. Their density is considerable, as much as 1,000 kg/ccm.

Zodiac

An imaginary belt on the celestial sphere along the ecliptic, encompassing the apparent paths of the Sun, the Moon and the planets. It includes twelve zodiacal constellations.

NOTE ON ILLUSTRATIONS

In contrast to current astronomical literature, our small guide does not contain any original photographs, but exclusively the drawings of celestial objects. The photorealistic drawings made by the author are based, naturally, on photographs – mostly on several photos of the same object taken by various observatories, with different instruments, exposure times, etc. The resulting drawing, consequently, represents a synthesis: one picture makes visible both very bright and very faint details of the same object. The drawing technique has made it possible also to prepare the individual "portraits" of celestial objects for print by the accentuation of very faint parts or colors of nebulae and galaxies. In this way, the illustrations have become clearer as it is required by popular science literature. Like every drawing, naturally, they may differ from genuine photographs, which are irreplaceable as the source of objective information for scientific purposes.

GENERAL INDEX

Air Pump *see* Antlia
Altar *see* Ara
amateur astronomer/astronomy 24, 28, 34, 36, 72, 96, 120, 170, 200
Andromeda 30, 31, 36, **46–47**, 112
angular measure/measurement 34, 48, 50, 52 ff., 218
Antlia (Air Pump) **48–49**
Apus (Bird of Paradise) **50–51**
Aquarius (Water Carrier) 17, 36, **52–53**
Aquila (Eagle) 42, **54–55**
Ara (Altar) 42, **56–57**
Aratus of Soli 7
Archer *see* Sagittarius
Aries (Ram) 16, 17, 36, **58–59**
Arrow *see* Sagitta
asterism 204, 206
asteroid 96
astrologer 18
astronomical literature 18, 87, 96, 218, 200
astronomical unit (AU) 8, 218, 220
astronomy, modern 28, 50, 64
astrophotography 174
atmosphere, Earth's 18, 74, 76, 96
–, stellar 29, 188
atmospheric extinction 74
atmospheric refraction 76, 88
Auriga (Charioteer) 11, 38, **60–61**
autumn constellations **36–37**
autumn equinox 138

Balance *or* Scales *see* Libra
Bartsch, Jacob 66, 150
Bayer, Johann 8, 50, 88, 108, 120, 128, 130, 152, 168, 200, 202, 212
Bayer's letter 32, 134
Berenice's Hair *see* Coma Berenices
Bessel, F.W. 104
Big Dipper 32, 34, 36, 40, 50, 62 ff.
binary *see* double star
binary constellation 42
binoculars 18, 28, 46 ff., 68, 73, 88, 120, 153
Bird of Paradise *see* Apus
black hole 152, 210, 218
Boötes (Herdsman) 40, 42, **62–63**
Brahe, Tycho 80
Bull *see* Taurus
Burnham, J.W. 106

Cacciatore, Niccolo/Nicolaus Venator 106
Caelum (Chisel) **64–65**
Camelopardalis (Giraffe) 34, **66–67**
Cancer (Crab) 17, 38, **68–69**
Canes Venatici (Hunting Dogs) **70–71**
Canis Major (Great Dog) 31, **72–73**
Canis Minor (Little Dog) 38, **74–75**
Capricornus (Goat) 17, 42, **76–77**
Carina (Keel) 36, 38, 40, 44, **78–79**, 102
Cassiopeia 9, 34,–39, **80–81**
"celestial fireworks" 92
celestial/stellar sphere 9, 10 ff.
Centaurus (Centaur) 40, 44, **82–83**, 102
central star 29, 48, 53, 54, 110, 121, 214
cepheid 24, 54, 84, 108, 162
Cepheus 34, 62, **84–85**
Cetus (Whale) 36, **86–87**
Chamaeleon (Chameleon) **88–89**
Charioteer *see* Auriga
Chisel *see* Caelum
Circinus (Compasses) **90–91**
circumpolar star 12, 34
Clark's refractor 106
cloud of interstellar dust and gas 26, 28, 29, 57, 61, 71, 102, 211;
see also Magellanic Cloud(s)
Colt *or* Foal *see* Equuleus
Columba (Dove) **92–93**
Coma Berenices (Berenice's Hair) **94–95**, 210
comet 18, 96
compass 12
Compasses *see* Circinus
Corona Australis (Southern Crown) 9, 11, 40, 42, 44, **96–97**
Corona Borealis (Northern Crown) 42, **98–99**, 210
Corvus (Crow) 40, **100–101**
Crab *see* Cancer
Crane *see* Grus
Crater (Cup) 40, **100–101**
Crow *see* Corvus
Crux (Southern Cross) 9, 11, 40, 44, **102–103**, 112
Cup *see* Crater
Cygnus (Swan) 42, **104–105**, 106

declination 10, 11, 14, 58
deep-sky/space objects 20–32, 64, 74, 88, 132, 158 f., 218
Delphinus (Dolphin) 42, **106–107**, 112
discovering new stars 92
Dorado (Swordfish) 44, **108–109**
double star *or* binary 24, 51 ff., 74, 92, 100, 110, 122, 130, 149, 218
–, close 110
–, double 145, 185
–, eclipsing 24, 60, 98, 144, 219
–, optical 24, 48–49, 76, 89, 98, 120, 218
–, physical 24, 218
–, spectroscopic 24, 118, 185
–, visual 24, 47, 91, 219
Dove *see* Columba
Draco (Dragon) 34, 42, **110–111**
dust ring 83
dwarf [star] 22, 53, 186; *see also* red dwarf; white dwarf

Eagle *see* Aquila
Earth 7–10, 12–18, 23, 58, 112
ecliptic 16, 17, 18, 40, 58, 59, 156
equator, celestial 10, 11, 13, 52, 58
equatorial constellations 13
equatorial coordinates 14, 34, 58
Equuleus (Colt *or* Foal) **112–113**
Eridanus *or* River Eridanus 36, 38, 44, **114–115**, 116, 124
evening star 18

"false cross" 40, 44, 78, 102, 156, 209, 212
First Point of Aries *see* vernal point
Fishes *see* Pisces
Flamsteed numbers 32, 134, 195
Fly *see* Musca
Flying Fish *see* Volans
Foal *or* Colt *see* Equuleus
Fornax (Furnace) 36, **116–117**
Fox *see* Vulpecula
Furnace *see* Fornax

galactic cluster 30, 31, 40, 94, 98, 99, 116, 124, 210, 218
"galactic windows" 26
Galaxy 26–28, 31, 44, 82, 99, 108, 116, 142, 158, 180, 182, 198, 202, 203, 219;
see also Milky Way
galaxy 30, 31, 40, 44, 88, 111, 116, 124, 130, 132 ff.
–, Andromeda *or* Great Andromeda 30, 46, 80, 108
–, Antennae *or* Ring-Tailed 100
–, Black-eye 94
: cluster of galaxies *see* galactic cluster
: colliding galaxies 69 f.
–, dwarf 116, 186
–, elliptical 30, 83
–, Fornax System 116
: Hubble's classification of galaxies 30
: Local Group of galaxies 30, 46, 66, 84, 116, 198
–, "satellite" 71
–, Sculptor System 186
: Seyfert galaxies 86, 93
–, Sombrero 211
–, Spindle 192
–, spiral 30, 46, 55, 66, 67, 71, 84, 86, 94, 114, 116, 124
– –, barred 67, 116, 124, 125
–, Stephan's Quintet 164
–, Whirlpool *or* Pinwheel 70
Gemini (Twins) 11, 17, 38, **118–119**
giant [star] 22, 23, 25, 54, 58, 60, 88, 100, 110, 126, 136, 203, 208
Giraffe *see* Camelopardalis
Goat *see* Capricornus
Great Bear *see* Ursa Major
Great Dog *see* Canis Major
"Great Square" of Pegasus 36, 46, 164
Grus (Crane) 36, **120–121**

H-II region 29, 78, 219
Hadley, John 156
Hare *see* Lepus
Hercules 42, **122–123**
Herdsman *see* Boötes
Hertzsprung-Russell Diagram (HRD) 21, 22, 88
Hevelius, Johannes 8, 50, 70, 89, 96, 107, 132, 140, 142, 150, 192, 193, 214

Hind, John Russell 136
horizon/horizontal plane 10, 12, 16, 74, 88
Horologium (Pendulum Clock) 44, **124–125**
hour circles 13
Houtman, Frederick de 8, 128
Hunter *see* Orion
Hunting Dogs *see* Canes Venatici
Huygens, Christian 124
Hyades 194, 195, 204
Hydra (Water Snake) 40, 48, **126–127**
Hydrus (Lesser Water Snake) **128–129**

Index Catalogue (IC) 32, 66
Indus (Indian) **130–131**
International Astronomical Union (IAU) 8, 64
interstellar matter *see* cloud of ...

Jupiter 18, 120

Keel *see* Carina
Kepler, Johann 66
Keyser, Pieter Dirkcszoon 8, 50, 88, 108, 128, 130, 162, 200, 202, 212

Lacaille *or* la Caille, Nicolas-Louis 8, 48, 64, 90, 116, 124, 146, 148, 152, 156, 170, 176, 178, 196
Lacerta (Lizard) **132–133**
latitude, geographical 9, 10, 11, 34, 88 ff.
Leo (Lion) 11, 17, 40, **134–135**
Leo Minor (Lesser Lion) **134–135**
Lepus (Hare) **136–137**
Lesser Water Snake *see* Hydrus
Libra (Scales *or* Balance) 17, 42, **138–139**
light pollution 74
light-year 20 ff., 219
Lion *see* Leo
Little Bear *see* Ursa Minor
Little Dog *see* Canis Minor
Lizard *see* Lacerta
Lupus (Wolf) 42, **140–141**
Lynx 34, **142–143**
Lyra (Lyre) 42, **144–145**

Magellanic Cloud(s) 44, 129, 146, 147, 157
–, Large 44, 108, 109, 146, 171, 180, 212
–, Small 44, 108, 146, 202, 203
magnitude 18, 20, ff., 220
–, absolute 20
–, apparent 18, 20, 56, 112, 124
–, changing 85
map, celestial *or* stellar 9, 16, 18, 32 ff., 140
– –, constellation 28, 32, **46–215**
– –, general **33–45**
–, pictorial 7, 126, 138, 182, 194
Mariner's Compass *see* Pyxis
Mars 18
Mensa *or* Table Mountain 109, **146–147**
Mercury 18

meridian, celestial 8, 12, 13
Messier, Charles 71, 132
Messier's catalogue (M) 32, 66, 132, 218
Messier's objects 132, 172, 182, 184
meteor 96
Microscopium (Microscope) 36, **148–149**
Milky Way 26–30, 32, 34, 38, 40, 42, 78, 102, 124, 153, 188; *see also* Galaxy
Monoceros (Unicorn) **150–151**
Moon 52, 58, 60, 68, 120
morning star 18
multiple star 24, 106, 119, 138, 204
Musca (Fly) 44, **152–153**

nebula 26, 29, 32, 220
–, bright 29,
–, dark *or* absorption 29, 44, 78, 102, 153
–, diffuse 29, 104, 109, 150
–, emission 29, 56, 57
–, planetary 29, 48, 52, 54, 88, 110, 121, 144, 214, 220
–, reflection 29
nebula, (named)
–, California 166
–, Cat's Eye 110
–, Coal Sack 44, 102, 153
–, Cone 151
–, Crab 132, 194, 196, 197, 221
–, Dumbbell 214
–, Eagle 190
–, Eskimo *or* Clown Face 119
–, Great (in Orion) 109
–, Helix 52
–, Horsehead 57, 161
–, Keyhole 78
–, Lagoon 182
–, North America 104
–, Omega *or* Horseshoe 182
–, Owl 205
–, Ring 144
–, Rosette 150, 153
–, Saturn 53
–, Tarantula 108, 109
–, Trifid 182, 183
nebulae, Veil 106
Neptune 18
neutron star 194, 208, 220
New General catalogue (NGC) 32
Norma (Rule) **154–155**
north celestial pole 10, 12, 34, 110, 207
north galactic pole 95
north pole of ecliptic 110
Northern Crown *see* Corona Borealis
northern sky constellations **33–43**
nova 24, 92, 96, 133, 152, 220
–, recurrent 92, 98, 176

observation, astronomical 9 ff., 12, 16, 28, 34, 50, 74, 88, 96, 112, 120, 178
Octans (Octant) 44, **156–157**
Ophiuchus (Serpent Bearer) 17, 42, **158–159**

Orion *or* Hunter 9, 13, 15, 38, 72, 74, **160–161**
Orion's belt 38, 72

Painter *see* Pictor
parallax 20, 104, 220
parallel, celestial 8
Pavo (Peacock) 44, **162–163**
Pegasus 36, 52, 112, **164–165**
Pendulum Clock *see* Horologium
Perseus 34, 36, 38, **166–167**
Phoenix 36, 44, **168–169**
photographing 174
Piazzi, Giuseppe 106
Pictor (Painter) 44, **170–171**
Pisces (Fishes) 17, 36, 58, **172–173**
Piscis Austrinus (Southern Fish) 36, 44, **174–175**
planet 18, 96
planetarium 9, 40, 96
planetoid 18
Platonic Year 58
Pleiades 38, 194, 195
Pluto 18
"Pointers" 44, 82, 102, 200
precession 58, 76, 110, 207
Ptolemy 7, 96, 112
pulsar 194, 208, 220
Puppis (Stern) **176–177**
Pyxis (Mariner's Compass) **176–177**

quadruple star 155

radiation 28, 30, 48, 57, 64, 74, 80, 86 ff., 218
radio astronomy 64
Ram *see* Aries
red dwarf 23, 82, 114 f.
Reticulum (Reticle) 44, **178–179**
right ascension 14, 16, 58
River Eridanus *see* Eridanus
Rosse, Lord 71
Rule *see* Norma

Sagitta (Arrow) 42, **180–181**
Sagittarius (Archer) 17, 42, **182–183**
Sails *see* Vela
satellites of the Earth, artificial 96
satellites of our Galaxy 108
Saturn 18
Scales *or* Balance *see* Libra
scintillation 18
Scorpius (Scorpion) 9, 17, 42, 56, **184–185**
Sculptor 36, **186–187**
Scutum (Shield) **188–189**
Serpens (Serpent) **190–191**
Serpens Caput (Serpent's Head) 42, 190
Serpens Cauda (Serpent's Tail) 42, 190
Serpent Bearer *see* Ophiuchus
Serpent's Tail *see* Serpens Cauda
Serpent's Head *see* Serpens Caput
Sextans (Sextant) **192–193**
Shield *see* Scutum
"smoker" 98

solar system 18, 86 ff.
south celestial pole 12, 44, 88, 156
south galactic pole 116, 186
Southern Cross *see* Crux
Southern Crown *see* Corona Australis
Southern Fish *see* Piscis Austrinus
southern sky constellations 33, **44–45**, 50 f., 102
Southern Triangle *see* Triangulum Australe
sphere, celestial 9 ff., 88, 112
spring constellations **41–42**
spring triangle 40, 54, 134
star 220–221
– brightness 20, 21, 24, 54, 84, 86, 108, 212
– chemical composition 22
– color 21, 22, 49, 64, 84, 90, 105
– definition 20
– diameter/size 20, 21, 22
– distance 20, 34, 84, 112
– evolution 21, 24
– formation 29, 100
– luminosity 20, 21, 84, 219, 220
– magnitude 20, 220
– main sequence 22
– mass 20, 104
– number 21, 88
– proper motion 62, 81,
– radiation, thermal 29, 90
– spectrum/spectral classes 21, 22, 50, 64, 88, 105
– subclasses 22
– temperature 21, 22, 24, 48 ff., 218
– types 22; *see main items alphabetically*
– wavelength 22
star, (named)
–, Acamar 114
–, Achernar 36, 38, 114, 124
–, Acrux 103
–, Acubens 68
–, Alamak 36, 46, 112
–, Albireo 105
–, Alchita *or* Al Chiba 100
–, Alcor 204
–, Aldebaran 38, 194
–, Algenib *or* Mirfak 166
–, Algieba 134
–, Algiedi 76
–, Algol 25, 36, 166
–, Algorab 100
–, Alkes 100
–, Almaaz 60
–, Alnair 120
–, Alphard 126
–, Al Risha 173
–, Alshain 54
–, Altair 42, 54
–, Antares 42, 138, 184
–, Archenar 36, 38, 44
–, Arcturus 40, 62
–, Arneb 136
–, Arrakis 110
–, Asellus Austrlis 68
–, Asellus Borealis 68
–, Barnard's 158
–, Betelgeuse 38, 74, 160
–, Canopus 36, 38, 44, 65, 78
–, Capella 38, 60
–, Castor 38, 118, 119
–, Cor Caroli 70
–, Deneb 42, 54, 104
–, Dubhe 34, 40
–, Enif 164
–, Fomalhaut 36, 44, 120, 174
–, Garnet *or* Erakis 84
–, Gemma 42, 98
–, Giedi 76
–, Hamal 58
–, Hind's Crimson 136
–, Izar *or* Pulcherrima 62
–, Kaffaljidmah 86
–, Kapteyn's 158, 171
–, Keid 114
–, Kepler's 158
–, Kitalpha 112
–, La Superba 70
–, Marfak *or* Marsic 122
–, Merak 34, 40
–, Mesarthim 58
–, Mira 86, 87
–, Mirach 36, 46
–, Mizar 204, 205
–, Peacock 44, 162
–, Polaris *or* North Star 34, 44, 66, 80, 110, 206, 207
–, Pollux 38, 118
–, Procyon 38, 74
–, Proxima Centauri 82
–, Ras Algethi 122
–, Rastaban 110
–, Regulus 40, 134
–, Rigel 38, 160
–, Rigil Kentaurus 82
–, Rotanev 106
–, Sadalmelik 52
–, Sadalsuud 52
–, Scheat 164
–, Sheliak 144
–, Sheratan 58
–, Sigma Scorpii 184
–, Sirius 18, 20, 38, 72, 74
–, Sirrah 36, 46, 112
–, Spica 20, 40, 100, 138
–, Struve 66
–, Sualocin 106
–, Thuban 110
–, Tycho's 80
–, Van Maanen's 172
–, Vega 20, 42, 54, 110, 144
–, Wolf-Rayet 208
star cluster 26, 28 f., 202, 221
– –, Butterfly 185
– –, Christmas Tree 151
– –, double 34, 166, 167
– –, globular 26, 28, 52, 57, 70, 76, 82, 96, 106, 137, 162
– –, Intergalactic Tramp 142
– –, Jewel Box 102, 103
– –, moving 38, 194; *see also* Hyades; Pleiades
– –, open 26, 28, 34, 55, 56, 60, 68, 72, 102, 105, 118
– –, Praesepe *or* Beehive 68
star companion 64, 66, 68, 72, 74, 80, 100, 110, 120
– –, Coathanger 214, 215
star group 214–215
"starting point" 9, 17, 44
Stern *see* Puppis
summer constellations **42–43**
summer triangle 42, 54, 104
Sun 12, 13, 16, 17, 20–23, 26 ff., 72, 74, 104, 112, 114, 120 ff.
supergiant 22, 23, 52, 66, 78, 86, 102, 104, 110, 136, 219
supernova 24, 80, 92, 106, 221
– remnant 9 2, 194, 196, 208
Swan *see* Cygnus
Swordfish *see* Dorado

Table Mountain *see* Mensa
Taurus (Bull) 16, 17, 38, **194–195**
telescope 18, 24, 28, 46 ff., 60, 64, 88, 103, 112, 148, 153, 190, 205
Telescopium (Telescope) **196–197**
temperature-luminosity *see* Hertzsprung-Russell Diagram
time, sidereal 13, 14, 15, 16
–, universal 178
Toucan *see* Tucana
Triangulum (Triangle) 36, **198–199**
Triangulum Australe (Southern Triangle) 44, 50, **200–201**
triple star 63, 68 f., 82, 112, 114, 151
Tucana (Toucan) 130, **202–203**
Twins *see* Gemini

Unicorn *see* Monoceros
Universe 31, 40, 99, 112
Uranus 18
Ursa Major (Great Bear) 9, 11, 31, 34, 40, 66, 110, **204–205**
Ursa Minor (Little Bear) 34, 110, **206–207**

variable star 24, 70, 78, 80, 84, 86, 98, 126, 136, 221
– light curve 24, 84, 98, 219
–, pulsating *see* cepheid
Vela (Sails) 38, 40, 44, 48, 102, **208–209**
Venus 18
vernal equinox *or* First Point of Aries 13, 17, 58, 59, 173
Virgo (Virgin) 17, 40, 138, **210–211**
Volans (Flying Fish) **212–213**
Vulpecula (Fox) **214–215**

Water Carrier *see* Aquarius
Water Snake *see* Hydra
Whale *see* Cetus
white dwarf 22, 23, 72, 74, 92, 114, 221
winter constellations **38–39**
winter hexagon 38, 60
winter triangle 38, 74, 150
Wolf *see* Lupus

X-ray (radiation) 82, 152, 210

zenith 9, 10, 12, 74, 76
zodiac 16, 221
–, signs of 16–18, 58, 68
–, constellations of 16–18, 52, 76, 221
Zwicky's catalogue 98